Basic Statistics

A Step by Step Guide

F. Devine

Harrap London

First published in Great Britain 1981
by Harrap Limited
19–23 Ludgate Hill, London EC4M 7PD

Reprinted 1982

© F. Devine 1981

ISBN 0 245-53495-4

Typeset by Western Printing Services Ltd. Bristol
Printed in Great Britain by Mackays of Chatham

Acknowledgments

This book is intended as a step by step guide to the basic techniques required in the statistical analysis of data. It attempts to meet the requirements of
    (*a*) the statistics options of mathematics courses at C.S.E. level,
    (*b*) full statistics courses at C.S.E. level,
and to provide a sound basis for pupils taking statistics at 'O' level.
I wish to thank Mrs. E. McConville, Mrs. J. Simm, Mr. A. S. Wilson, Mr. J. Hughes and Mrs. A. M. Devine for their invaluable help during the preparation of this book.
I am indebted to the pupils of Ellergreen Comprehensive School, Liverpool, on whom the original draft of the book was tried and tested over a five year period. The successes they achieved at C.S.E. level persuaded the author that the style and approach adopted throughout the book was useful and reassuring, especially to the average and below average pupil.
I am thankful to the following examination boards who have granted permission to use their past examination questions as exercises.

East Midland Regional Examinations Board.	E.M.
Yorkshire Regional Examinations Board.	Y.
Associated Lancashire Schools Examining Board.	L.
Northern Ireland C.S.E. Examinations Board.	N.I.
East Anglian Examinations Board.	E.A.
West Midlands Examinations Board.	W.M.
Middlesex Regional Examining Board.	M.
North West Regional Examinations Board.	N.W.

F. Devine
September, 1980.

Contents

1 Collection of Data 1

2 Tabulation of Data 8

3 Diagrams 34

4 Averages 75

5 Dispersion 115

6 Probability I 148

7 Correlation I Scatter Diagrams 174

8 Moving Averages 189

9 Correlation II Ranks 202

10 Use of Guessed Mean 212

11 Weighted Averages 237

12 Probability II 248

13 Sampling Techniques 261

Appendix 269

1 COLLECTION OF DATA

Origins

Statistics began as the study of states and countries.
Such quantities as the:

(a) wealth   (c) industrial output
(b) military strength   (d) agricultural output,

of each state were calculated and then compared with those in other states.
Only quantities of interest to the state were studied in detail.

Today

In the modern world, statistics has become the study of any kind of data,
i.e. facts, which can be of help to society.
Statistics are used in every walk of life; they affect all of us, whether we
realise it or not.
Some of the many areas where statistics are used include advertising, stock
control, sales trends, market research, Football League tables, T.V. ratings
figures, banking, insurances and climatology.

Facts (Data)

Facts are often required by various people so that they can make decisions to:

(a) leave things as they are
(b) change things to make them better.

Examples: The Government may want to know how many people there are
in the country:

(a) altogether   (d) who are single   (g) living in council houses
(b) working   (e) who are married   (h) living in private houses.
(c) unemployed   (f) of each sex

Washing powder manufacturers may want to know:

(a) how many people use their brand
(b) how many people use brand X.

Town planners may want to know if a new road will:

(*a*) reduce traffic jams
(*b*) increase accidents
(*c*) cause damage to houses.

A light bulb manufacturer would want to know how many hours his light bulbs last.

Population

Data is collected about many populations.
A statistical population does not have to consist of a set of people.
A statistical population consists of all of the items about which data is being collected.
The following table gives some examples of statistical populations.

Data being collected	*Population*
The life of a particular type of light bulb.	All the light bulbs of that type.
The variability in the weights of sugar bags filled on a production line.	All the sugar bags filled on that production line.
The ages of people in Great Britain.	All the people in Great Britain.
The colours of hair of pupils in a class.	All the pupils in that class.
The eating habits of a colony of ants.	All the ants in that colony.
The number of words per line in a mathematics textbook.	All the lines in that textbook.
The number of letters per word in the Bible.	All the words in the Bible.
The number of eggs laid by a battery of hens.	All the hens in that battery.

When collecting data about a population, it is desirable:

(*a*) to collect data about every member of the population.
(*b*) to collect the data required as fast as possible.
(*c*) to keep the cost of collection as low as possible.
(*d*) not to destroy the whole population.

If a population is small, then data can be collected about every member of the population quickly and cheaply.

However:

(a) most populations are large or even infinite (unlimited) and collection of data about every member is slow and expensive.
(b) in many cases collecting data about a whole population will destroy the population. E.g. if all light bulbs of a certain type were tested to find out the life of that type of bulb, there would be no light bulbs left to sell!

If a population is large or if data collection from the whole population would lead to its destruction, it is usual to collect data about the population by using a representative sample of the population.

Sampling

A sample is made up of some of the members of a population.
A representative or good sample is one in which the results obtained for the sample can be taken to be true for the whole population.

Examples:

(a) If 30 % of a good sample smoke then we can say that 30% of the population smoke.
(b) If 60% of a good sample watch Coronation Street then we can say that 60% of the population watch Coronation Street.
(c) If 98% of a good sample of light bulbs last at least 1000 hours then we can say that 98% of light bulbs of that type will last at least 1000 hours.

To be a representative or good sample, the members of the sample must be chosen carefully.
A good sample must be chosen:

(a) at random, i.e. every member of the population must have a chance of being chosen.
(b) so that it is large enough to satisfy the needs of the investigation being undertaken.
(c) so that it is in no way biased.

Choosing at random does not mean choosing haphazardly i.e. any old how, but choosing as in a raffle, by giving every member of the population a number and then picking numbers from a hat.
For large populations numbers are not taken from a hat, but are taken from

tables of random numbers or are created by a random number generating machine e.g. a computer.

A sample must be large enough to ensure that if some of the extreme members, i.e. very high or very low members of a population, are chosen to be in the sample, they do not affect any results obtained from the sample more than they should.

The larger the sample chosen, the better it will represent the population as a whole.

When is a sample large enough? The answer to this question depends on the size of the population.

A sample with 10 members may be large enough or not, depending on the size of the population.

A sample of 10 out of a population of 30 is a large sample, e.g. 10 pupils out of a class of 30 pupils if they are being asked their views on how well a subject is taught to them.

A sample of 10 out of a population of 1500 is a small sample, e.g. the views of 10 pupils out of a school of 1500 pupils on how well the school is run.

Bias (Favour)

Bias will occur if samples are chosen:

   (*a*) deliberately by a person as this will lead to favouritism.
   (*b*) in a haphazard fashion.

A sample from a population which has been chosen to be:

   (*a*) random   (*b*) large enough

should turn out to be completely unbiased.

An unbiased sample is one which will give the views of each section of a population in a balanced way, i.e. each section will have its fair say.

An unbiased sample should be made up of members from each section of the population.

The number of members from each section that should be present in the sample depends on the size of that section.

The bigger the section is, the more members it should have in the sample.

No section should have either more or less than its fair number of members in the sample.

A sample contains some fraction or percentage of the population.

This *same* fraction or percentage of each section of the population should be found in the sample.

Example: If the views of the people who run a school are needed about a change in working hours, then the views of everyone concerned with running the school should be sought. If a sample of 1/5th or 20% of the population is considered large enough then the sample should contain 1/5th or 20% of:

(*a*) kitchen staff
(*b*) teaching staff
(*c*) office staff
(*d*) cleaning staff
(*e*) governors
(*f*) technicians.

A school with 20 kitchen staff, 60 teaching staff, 5 office staff, 10 cleaners, 15 governors and 5 technicians has a total population of interested parties of 115. A 20% sample of this population would have 23 members which should be made up of 4 kitchen staff, 12 teaching staff, 1 of the office staff, 2 cleaners, 3 governors and 1 technician, i.e. 1/5th of 20, 60 etc.

All samples should be checked with care to ensure that they are free from bias or favour.

A random sample which gave 2 kitchen staff, 16 teaching staff, 1 office staff, 2 cleaners, 1 governor and 1 technician would not be balanced. It is biased towards (favours) the teaching staff and is biased against the kitchen staff and governors.

Even when the random sample contains the correct number of members taken from each section of the population, bias can still occur by:

(*a*) substituting for a member who was chosen at random to be in the sample with another member. E.g. if a cleaner chosen is òff sick, choosing another cleaner instead.
(*b*) failing to complete the survey for the whole sample. E.g. by not bothering to contact the sick cleaner at home.

Data Collection

Data can be collected by:

(*a*) questioning people and noting their replies,
(*b*) watching and noting what happens in a given situation,
(*c*) doing experiments,
(*d*) looking up data already collected by other people.

Note: Chapter 13 on page 261 deals in greater detail with data collection, questionnaires and sampling methods.
The appendix on page 268 gives some ideas for data collection projects.

Exercise 1: Collection of data from different sized populations.

1. *Population A:* yourself.

Find and write down your:

(*a*) age to the nearest month,   (*f*) weight to the nearest kilogram,
(*b*) height to the nearest centimetre,   (*g*) shoe size,
(*c*) waist size to the nearest centimetre, (*h*) colour of hair,
(*d*) chest size to the nearest centimetre, (*i*) colour of eyes.
(*e*) hip size to the nearest centimetre,

Do colours cause a problem? Discuss this as a class with your teacher.

2. *Population B:* your family.

Find out and write down for each member of your family:

(*a*) their favourite breakfast food,
(*b*) their favourite drink,
(*c*) their favourite colour,
(*d*) the colour of their eyes,
(*e*) their ages,
(*f*) at least five other items of interest about each member of your family.

Do not forget that this population includes yourself.

3. *Population C:* your class.

Make a table similar to the one below and fill it in for the whole of your class.

Name	Age	Height	Waist	Chest	Hips	Weight	Shoe Size	Hair	Eyes
Alison									
Amanda									
Barbara									

4. *Population D:* your year group.

Discuss as a class how to collect information on population D.
Is it better to work on your own or as a class?
Is it possible to collect information from every member of the population?
If you decide the answer is yes, discuss and decide as a class on the best method of collecting the information. If the answer is no, discuss and

decide how the difficulties in collecting information from a large population can be overcome.

Collect the following information from the pupils in your year group:

(a) their favourite hobbies.
(b) their favourite pop star.
(c) their favourite T.V. programme.
(d) their favourite football team.
(e) the month in which their birthday falls.

5. *Population E:* your school.

Discuss as a class how to collect information on population E.
Do you need to choose a sample?
If so how would you select the sample?
Collect the following information from the pupils in your school:

(a) favourite subject.
(b) least popular subject.
(c) pets kept.
(d) intended career after leaving school.

6. As an individual pick one of the data collection projects in the appendix on pages 268–9 and using a suitable population collect the required data.

7. As a class design a data collection project and using a suitable population collect any required data.

8. Conduct a survey of the use of space in one of the following:

(a) an office block,
(b) a shopping precinct,
(c) an industrial estate,
(d) a market.

9. Conduct a survey of:

(a) the types of buildings in your area,
(b) the amenities in your area.

2 TABULATION OF DATA

Processing Data

Data collected during a survey is called raw or crude data.
Raw data as it stands is not usually of much use to the surveyors.
Raw data can be 'refined' or processed to make it more useful.
The raw data can be 'refined' in a variety of ways.
Data which has been 'refined' can be further 'refined' and so on: i.e. raw data can undergo multi-stage refinement.

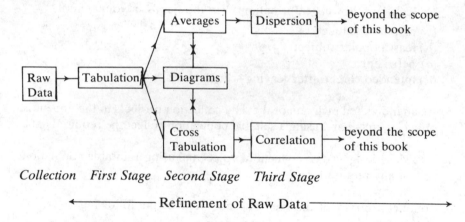

Collection   First Stage   Second Stage   Third Stage

← ——————————— Refinement of Raw Data ——————————— →

The above diagram is an illustration of how raw data might be refined.
The primary, or first stage, in the refinement of data, is normally to arrange it into tables which show the results of a survey.
Making tables of raw data is called *tabulating* the data.

The types of tables which we will consider are:

    1. array   3. frequency distribution
    2. tally chart   4. cumulative frequency distribution.

When tabulating data it is very important to:

1. be neat
2. be accurate
3. make any tables as simple as possible to understand
4. make sure that any titles and headings can have only one possible meaning.

Tabulation

1. Array
Rows of data go across the page.
Columns of data go down the page.
When data is arranged so that it appears as both rows and columns in the same table, the table is called an *array*.
An array should have distinct columns and distinct rows.

Example: The following array shows the results of a survey to find the colours of socks worn by the staff at a school.

Green	Blue	Red	Green	White	Blue	Yellow	←*row*
Red	Green	White	Blue	Red	Red	Blue	
Black	Red	Green	Maroon	Red	Yellow	Red	
Blue	White	Maroon	Black	Red	Red	Red	
Green	Yellow	Blue	Blue	Red	Black	Red	

↑
Column

The above array has 5 rows and 7 columns.
This array is called a 5 × 7 array.

2. Tally Chart
An array shows all the data in a neat table but does not tell us immediately how many times each item in the array turned up. To find out how often each item occurred, a tally chart can be constructed.
To construct a tally chart:

    (*a*) make a list of all the results possible in the data, or in an array, down the left hand side of a piece of paper.
    (*b*) go through the data one item at a time and place a *tally* (stroke) against the correct result in the list, each time that result turns up.

It is important when making a tally chart to:

1. enter *only one* tally at a time.
2. cross out each item in an array as soon as you have entered it on the tally chart. This ensures that you do not make mistakes by counting something twice.
3. enter each tally against the correct result.
4. make sure that all of the data is crossed out at the end.
5. count the tallies at the end to make sure that the total number of tallies is the same as the number of items of data you began with.

It is usual to mark tallies in groups of five. When a row of four tallies has been obtained thus, IIII, the next tally is placed across the others thus, IIII, to make a group of five tallies. The reason for doing this is that when counting the tallies at the end it is easier to count in fives.

The fastest and most efficient way of tallying an array is to tally each item on the first row, one after the other, in the order that they appear in the row, so that all of the row is crossed out one by one. Then repeat this for the 2nd row and so on. This method of tallying is much faster than going through the array looking for all the items of a particular type, then going through again and again for items of other types.

Example: Make a tally chart of the colours shown in the array of the previous example about Staff socks.

Colours	Tallies
Green	IIII
Blue	IIII II
Red	IIII IIII II
White	III
Black	III
Maroon	II
Yellow	III

Array had 5 rows × 7 columns giving 35 colours.
Tally chart has 35 tallies.

Example: The table (array) below, shows the marks obtained in a test by a class of 40 pupils. Make a tally chart of these marks.

```
8  2  1  5  7  6  4  3  7  8
9  7  5  4  6  6  1  3  9  5
8 10  8  9  2  1  6  3  1  4
0  6  8  4  4  1 10  9  1  3
```

4 rows × 10 columns = 40 marks.

Mark	Tallies
0	I
1	~~IIII~~ I
2	II
3	IIII
4	~~IIII~~
5	III
6	~~IIII~~
7	III
8	~~IIII~~
9	IIII
10	II

Tally chart has 40 tallies.
Array had 4 rows × 10 columns
giving 40 marks.

3. Frequency Distribution Table

When a tally chart has been constructed, it is usual to put the total of the
tallies for each item in a column on the right hand side of the tally chart.
A tally chart with the totals included is called a *frequency distribution table*.
The frequency of an item is the number of times that item turns up.

Example: The frequency distribution for the colours of socks would be:

Colours	Tallies	Frequency Total
Green	~~IIII~~	5
Blue	~~IIII~~ II	7
Red	~~IIII~~ ~~IIII~~ II	12
White	III	3
Black	III	3
Maroon	II	2
Yellow	III	3

Number of times a
colour turned up.

35 ◄────── Grand Total

Example: The frequency distribution for the marks would be:

Mark	Tallies	Frequency Total
0	I	1
1	IIII I	6
2	II	2
3	IIII	4
4	IIII	5
5	III	3
6	IIII	5
7	III	3
8	IIII	5
9	IIII	4
10	II	2

Number of times a mark turned up.

40 ◄———— Grand Total

A *frequency distribution* does not have to have tallies.
A *frequency distribution must* have a list of items and the number of times that they turn up i.e. their frequencies.

Examples:

Colour	Frequency
Green	5
Blue	7
Red	12
White	3
Black	3
Maroon	2
Yellow	3

Mark	Frequency
0	1
1	6
2	2
3	4
4	5
5	3
6	5
7	3
8	5
9	4
10	2

Frequency distributions may be written *across* the page as well as down the page.

Examples:

Colour	Green	Blue	Red	White	Black	Maroon	Yellow
Frequency	5	7	12	3	3	2	3

Mark	0	1	2	3	4	5	6	7	8	9	10
Frequency	1	6	2	4	5	3	5	3	5	4	2

The above are frequency distributions which are going *across* the page.

4. Cumulative Frequency Distribution (Running Totals Table)

A frequency distribution is a table which shows how many times a particular value or item turned up in the original data.

A frequency distribution, as it stands, *will not* show easily how many values or items in the data are *less* than a particular value.

A frequency distribution, as it stands, *will not* show easily how many values or items in the data are *more* than a particular value.

E.g. a frequency distribution will show how many people obtained a mark of 6 but it will not show, without doing some working out, how many people obtained less than 6.

A frequency distribution can be changed into a
Cumulative frequency distribution.

A cumulative frequency distribution *will* show how many values there are:

(*a*) *more* than a particular value
(*b*) *less* than a particular value.

To change a frequency distribution into a cumulative frequency distribution to show *less than:*

(*a*) make two columns to the right of the frequency (totals) column.
(*b*) head the first new column *cumulative frequency.*
(*c*) head the second new column *less than or equal to*, or use the symbol ≤.
(*d*) fill in the less than or equal to column by simply repeating all of the values which appear in the first column of the frequency distribution.
(*e*) fill in the cumulative frequency column by:

1. taking the first number in the frequency or *F* column and making it the first number in the cumulative frequency or *CF* column.

2. adding this first number in the *CF* column to the second number in the *F* column and making the answer the second number in the *CF* column.

3. adding this second number in the *CF* column to the third number in the *F* column and making the answer the third number in the *CF* column.

4. repeating the process, working down the table, until all the *CF* column is filled in.

Example:

Frequency Distribution   Cumulative Frequency Distribution

Mark	Freq.
0	1
1	2
2	5
3	8
4	12
5	7
6	6
7	4
8	3
9	1
10	1

becomes →

Mark	Freq. F	Cum. Freq. CF	Less than or Equal to ≤
0	1	1	0
1	2	3	1
2	5	8	2
3	8	16	3
4	12	28	4
5	7	35	5
6	6	41	6
7	4	45	7
8	3	48	8
9	1	49	9
10	1	50	10

These are the same figures as those in the marks column.

50←These two numbers should be the same.

From the frequency distribution it can be seen that 7 pupils obtained 5 marks.

From the cumulative frequency distribution it can now be seen that 35 pupils obtained 5 or less than 5 marks.

From the cumulative frequency distribution it can also be seen that 28 pupils obtained less than 5 marks.

Example: Make a cumulative frequency distribution to show less than for the marks obtained by 120 pupils in a maths test. The results are shown in the following array.

```
2 1 3 2 4 4 3 5 4 3 4 3 4 2 3 3 7 7 3 5
3 5 4 5 4 5 4 6 5 4 5 5 4 4 2 4 4 5 5 5
5 6 5 6 3 7 6 5 6 2 6 5 5 3 5 5 4 6 6 4
3 4 5 5 5 4 5 4 5 7 4 4 3 2 4 6 3 6 5 3
5 6 6 3 8 6 3 6 6 3 9 7 6 4 3 4 6 3 3 5
4 5 4 5 4 8 5 4 4 6 4 6 4 4 4 7 5 5 4 2
```

Marks	Tallies	Freq.	Cum. Freq.	Less than or Equal to ≤
1	I	1	1	1
2	JHT II	7	8	2
3	JHT JHT JHT JHT	20	28	3
4	JHT JHT JHT JHT JHT JHT III	33	61	4
5	JHT JHT JHT JHT JHT JHT I	31	92	5
6	JHT JHT JHT IIII	19	111	6
7	JHT I	6	117	7
8	II	2	119	8
9	I	1	120	9

To change a frequency distribution into a cumulative frequency distribution to show *more than*:

(a) write out the frequency distribution upside down.

(b) using the upside down frequency distribution, follow the same procedure as that used for less than, but head the last column *more than or equal to*, or use the symbol ≥.

Example:

Mark	Freq.		Mark	Freq.		Mark	Freq.	Cum. Freq.	More than or Equal to ≥
1	3		9	1		9	1	1	9
2	5		8	3		8	3	4	8
3	10		7	5		7	5	9	7
4	16		6	7		6	7	16	6
5	20	→	5	20	→	5	20	36	5
6	7		4	16		4	16	52	4
7	5		3	10		3	10	62	3
8	3		2	5		2	5	67	2
9	1		1	3		1	3	70	1

From the frequency distribution it can be seen that 16 pupils obtained 4 marks.

From the cumulative frequency distribution it can be seen that 52 pupils obtained 4 or more than 4 marks.

From the cumulative frequency distribution it can also be seen that 36 pupils obtained more than 4 marks (that is the same as, more than, or equal to 5 marks).

Exercise 2a:

1. Make a frequency distribution of the following array.

```
1  3  7  3  6  0  5  2  6  2
6  1  6  2  6  3  6  5  3  3
7  8  5  9  7  7  9  8  6  8
```

2. Make a frequency distribution of the following array.

```
9  1  6  5  1  3  5  6  7  9  2  5  8  4  6
5  5  6  4  3  3  4  3  2  5  5  4  6  7  8
0  7  6  5  4  5  8  6  4  5  5  4  3  2  8
```

3. Make a frequency distribution of the following array.

```
0  4  2  4  4  4  5  7  4  4  5  3
1  5  7  1  0  7  6  1  3  4  3  5
6  3  8  1  3  2  2  0  4  4  3  7
8  9  6  3  4  2  1  4  2  5  4  4
6  3  4  3  2  2  5  3  4  3  4  2
```

4. Make a frequency distribution of the following array.

```
5  1  4  0  3  5  2  6  4  6
3  6  9  6  4  8  2  3  5  4
2  2  5  0  3  5  5  4  3  4
4  6  6  8  4  5  5  4  3  2
8  6  3  5  7  6  1  5  6  2
3  2  4  7  4  4  2  1  6  5
7  3  5  5  5  4  5  6  4  3
4  6  7  5  5  4  3  8  5  4
6  9  2  4  7  3  8  5  6  3
4  5  7  3  5  7  5  6  4  5
```

5. The following array shows the last letter of the registration numbers of the cars which passed an observer during a traffic survey. Make a frequency distribution of these registration letters.

```
S  R  P  R  T  A  R  N  S  S  R  M  K  R  K
N  T  M  L  K  P  K  P  M  B  P  N  T  S  N
J  H  R  J  R  F  T  L  P  N  R  H  P  H  P
P  F  P  S  L  P  H  T  L  S  J  N  S  R  N
D  S  H  N  L  R  R  P  T  P  N  M  L  J  N
C  D  N  N  L  R  R  P  N  R  N  M  T  S  M
S  R  J  T  L  N  R  P  N  R  R  S  P  E  M
S  C  P  K  R  M  P  R  T  N  K  L  R  N  R
R  P  S  M  C  S  N  S  N  P  T  R  K  L  E
R  E  M  R  T  P  M  R  K  S  T  R  E  S  L
```

6. Copy the following passage into your book, being careful to copy it exactly as it appears.

One of the most popular holiday resorts nestles between the sun-kissed, wave-washed, golden sands of the south and the awe inspiring snow-capped mountains of the north. The rich blue sea offers every type of water-based activity, while the gentle slopes leading to the mountains abound with green foliage and are rich with the fragrances given off by the abundance of fruits and herbs. These foothills are ideal for those who prefer to walk or pony trek. The night life caters for all tastes and age groups.

Make a frequency distribution of:

(*a*) the number of letters per word in the passage.
(*b*) the number of vowels per word in the passage.
(*c*) the vowels contained in the whole passage.
(*d*) the letters contained in the whole passage.

7. Use the table of data about your class which you constructed for question 3 of Exercise 1 (page 6) to make frequency distributions of the:

(*a*) ages,	(*e*) hip sizes,	(*h*) hair colorations,
(*b*) heights,	(*f*) weights,	(*i*) eye colorations,
(*c*) waist sizes,	(*g*) shoe sizes,	of the class.
(*d*) chest sizes,		

8. Construct a cumulative frequency distribution showing 'less than' of the array shown in question 1.

9. Construct a cumulative frequency distribution showing 'more than' of the array shown in question 1.

10. Construct a cumulative frequency distribution showing 'less than' of the array shown in question 2.

11. Construct a cumulative frequency distribution showing 'more than' of the array shown in question 2.

12. Construct a cumulative frequency distribution showing 'less than' of the array shown in question 3.

13. Construct a cumulative frequency distribution showing 'more than' of the array shown in question 3.

14. Construct a cumulative frequency distribution showing 'less than' of the array shown in question 4.

15. Construct a cumulative frequency distribution showing 'more than' of the array shown in question 4.

Grouping Data

Range
The range of a set of data is the difference between the highest value in the data and the lowest value.
To find the range of a set of data:

(*a*) find the highest value in the data.
(*b*) find the lowest value in the data.
(*c*) subtract the answer to (*b*) from the answer to (*a*).

When data is spread out over a wide range, any tables produced from the data, using the methods already described, i.e. with one line for each different value in the data, will be long and unmanageable.
It is sensible and very useful to arrange the data in groups or *class intervals*, with a number of different values placed in each class interval.
Grouping data in class intervals has the effect of shortening any tables produced from the data to easily manageable lengths.
E.g. the results of exams marked out of 100 are usually such that the marks obtained by pupils are spread out between 0 and 100.

Range = Highest value–Lowest value
The range in this case is $100-0 = 100$.

To make a tally chart with one line for each mark, would take 101 lines and there would be very few or no tallies on most of these lines.
Exam marks are usually grouped in classes with 10 marks in each class interval, e.g. class intervals of 0–9, 10–19 etc., might be used.

If marks are grouped together with 10 marks per class interval, then the table only needs 10 classes and so will only need 10 lines.

If the same exam marks were grouped in classes with 5 marks in each class interval, e.g. 0–4, 5–9, 10–14 etc., then the table would need 20 classes and so it would need 20 lines.

Data can be spread out over a wide range, but in such a way that most of the values are concentrated close to the middle of the range.

Spreading of this kind, results in the middle class intervals having a large number of values in each, while the intervals at both ends of the range have few or no values in them.

In cases like this it is usual to use small intervals for the middle classes and large intervals for the end classes.

Class intervals need not always be the same size.

E.g. 0–19, 20–29, 30–34, 35–39, 40–44, 45–49, 50–59, 60–69, 70–100.

In the above example there is an interval with 20 values. There are others with 10 values, others with 5 values, while the last interval has 31 values.

Class Limits

Every class interval has two stated limits, e.g. 20–29. The stated limits of this interval are 20 and 29.

The larger stated limit of an interval is the *upper class limit* of the interval.

The smaller stated limit of an interval is the *lower class limit* of the interval, e.g. the interval 20–29 has an upper class limit of 29 and a lower class limit of 20.

The interval 41–50 has an upper class limit of 50 and a lower class limit of 41.

Any value lying between the limits of a class is in that class.

Note: The final class, 70–100, of the example above could be written as 70 and over or as 70+. These two ways of writing the interval are examples of *open ended* intervals, since there are no stated upper limits.

Open ended class intervals are useful when tabulating data, but situations arise during the processing of the data when an interval must be *closed*, e.g. calculating the mean, drawing histograms.

An open ended class interval should be closed at a *sensible* limit, i.e. where the limit would most probably be.

E.g. an age interval of 65+ might be closed at 105, since very few people are likely to live beyond that age.

An interval of 80+ for exam marks would be closed at 100, since that is the highest mark possible if the exam is marked out of 100.

The *mid-value* of a class interval is the middle value of the interval.

If an interval has been chosen carefully, all of the values in the interval should be spread out around the mid-value of the interval.

The mid-value is often used to represent the class as a whole.
To find the mid-value of a class interval:

  (*a*) add the lower limit of the interval to the upper limit of the interval.
  (*b*) divide the answer to (*a*) by 2.

E.g. for 1–20 we have $1 + 20 = 21$ and $21 \div 2 = 10\frac{1}{2}$ so the mid-value $= 10\frac{1}{2}$.

Variables
Each member of a given population has a variety of characteristics, i.e. properties, qualities or attributes.
E.g. all people have height, weight, colour of hair, intelligence etc.
Any single characteristic will be possessed by each member of the population, e.g.
  All people have an age.
  All matchboxes have a contents, even if it is zero.
  All sugar bags have a weight.
The size of a characteristic may vary, i.e. be different, in each member of the population, e.g.
  People have different shoe sizes.
  Stars have different degrees of brightness.
  Earthquakes have different seismic intensities.
A variable is a characteristic which can vary in size.
Variables can be divided into two types:

  (*a*) discrete   (*b*) continuous

A *discrete variable* is one which can take up only definite fixed values. It cannot take any values between these definite values.
This means that there are gaps between those values which the variable can assume. The values which the variable can assume go up in steps or jumps.

Examples:
Wind strengths on the Beaufort Scale are measured as 0, 1, 2, 3, 4, 5, 6, 7, 8, 9, 10, 11, 12.
Eggs are graded in sizes 1, 2, 3, 4, 5, 6.
Shoes are measured in half sizes 3, $3\frac{1}{2}$, 4, $4\frac{1}{2}$, 5 etc.
Heights of planes are stated in Flight Levels 54, 55, 56 etc.
Postal charges for letters go up in jumps, depending on the weight of the letter.

A *continuous variable* is one which can take up any value within a given range. Since the variable can assume any value, there are no gaps. The values which the variable can assume go up continuously like the slope of a hill.

Examples:
Wind strengths can be measured as any speed between 0 and 100 mph (say)
e.g. 79.45 mph.
Eggs can have any weight between 50 and 80 g.
The length of a foot may have any length between 5 and 30 cm.
The actual height of a plane at Flight Level 55 may in fact be any height
between 5500 ft and 5600 ft.
The actual weight of a letter may be any weight between 3 g and 300 g.
When collecting data about a continuous variable, the size of the variable
must be measured for each member of a population or sample, e.g. heights,
weights.
The measurements of the size of a variable can be performed very
accurately, but before tabulation in an array, all measurements must be
rounded off to the necessary degree of accuracy, e.g. a weight measured as
54.2765 kg might be needed to the nearest kg giving a tabulated value of
54 kg. The same weight, but needed to one decimal place, would have a
tabulated value of 54.3 kg.

Exercise 2b:

1. A set of exam marks was grouped into class intervals of 0–9, 10–19,
   20–29, 30–39, 40–49, 50–59, 60–69, 70–79, 80–89, 90–100.

   (*a*) What was the greatest possible range of the marks?
   (*b*) What is the lower class limit of the 2nd class?
   (*c*) What is the upper class limit of the 1st class?
   (*d*) What is the lower class limit of the 6th class?
   (*e*) What is the upper class limit of the 8th class?
   (*f*) What are the mid-values of the 2nd, 5th, and 7th classes?
   (*g*) If the highest mark was 93 and the lowest mark was 5, what was the
   actual range of the marks?

2. In a survey of the pocket money given to the members of a class by their
   parents, the amounts were grouped into class intervals (in pence) of
   30–39, 40–59, 60–79, 80–109, 110–129, 130–139, 140–150.

   (*a*) What was the greatest possible range of pocket money given to the
   class?
   (*b*) What are the mid-values of each of the classes?
   (*c*) What are the upper class limits of the 1st, 3rd, 4th and 7th classes?
   (*d*) What are the lower class limits of the 1st, 4th, 5th and 7th classes?
   (*e*) Which class intervals have the same range?
   (*f*) Which class interval has the greatest range?

3. Use the data you collected about your class to find the range of the:

(*a*) ages,	(*e*) hip sizes,
(*b*) heights,	(*f*) weights,
(*c*) waist sizes,	(*g*) shoe sizes,
(*d*) chest sizes,	of the pupils in your class.

4. Repeat question 3 considering only the girls in your class.

5. Repeat question 3 considering only the boys in your class.

6. Which of the ranges in questions 4 and 5 would you expect to be different? Check the ranges to see if they are different.

7. State which of the following are continuous and which are discrete variables:

(*a*) number of roses per rosebush in a garden.
(*b*) heights of rosebushes in a garden.
(*c*) temperature measured throughout one day.
(*d*) volume of oil in the sump of a car each day over one year.
(*e*) volume of oil which can be bought at garages.
(*f*) rainfall per day throughout a week.
(*g*) the number of people employed at each local firm.
(*h*) weights of people employed by local firms.

8. Discuss with your teacher as many variables as you and your class can think of. Make a table showing the results of your discussion, with one column for continuous variables and another for discrete variables.

Class Boundaries

Every class interval has a lower class limit and an upper class limit, e.g.

The class intervals 1–10, 11–20, 21–30 all have limits.

The class 1–10 has an upper class limit of 10.

The class 11–20 has a lower class limit of 11.

When grouping continuous data, there may be values which fall between the upper class limit of one class and the lower class limit of the next class. E.g. 10.2, 10.5, 10.7, all fall between 10 and 11.

Which class to use for each of these values is a problem, as they lie between the limits of two classes.

This problem can be overcome if we consider each class to extend out from its limits as far as two boundaries, one on either side of the class.

Each class has two boundaries, an upper class boundary and a lower class boundary.

The boundaries between two class intervals meet halfway between the class limits of the intervals.

The upper class boundary of one interval is exactly the same as the lower class boundary of the next interval.

To find the boundary between two class intervals:

(a) add the upper limit of one class to the lower limit of the next class.

(b) divide the answer to (a) by 2.

E.g. for the class intervals 21–30 and 31–40, the upper limit of first class = 30, the lower limit of second class = 31.

$30 + 31 = 61$ and $61 \div 2 = 30\frac{1}{2}$, so the boundary between the classes is $30\frac{1}{2}$.

For the intervals ↑ 1–10 ↑ 11–20 ↑ 21–30 ↑ 31–40 ↑ 41–50 ↑
boundaries halfway between the limits

There is a boundary before the first class and another boundary after the last class.

The boundaries of the above class intervals are 0.5, 10.5, 20.5, 30.5, 40.5, 50.5.

The boundaries 0.5 and 50.5 are outside of their respective limits.

Both 0.5 and 50.5 are outside their limits by the same amount as the other boundaries in the table are outside their limits.

10.5 is the upper class boundary of the interval 1–10 and also the lower class boundary of the interval 11–20.

Any value lying within the boundaries of a class is in that class.

Using this idea of boundaries, the value 10.2 is obviously in the interval 1–10, and 10.7 is obviously in the interval 11–20, but 10.5 is right on the boundary between two classes.

The commonly accepted practice is to place any value on the boundary between two classes, in the higher interval, so 10.5 is placed in the interval 11–20.

It may be that when grouping continuous data it is useful to group the data so that the upper class limit of each class is the same as the lower class limit of the next class.

E.g. times in seconds could be grouped as follows:

0–10, 10–20, 20–30, 30–40, 40–50, 50–60 etc.

In such cases the class limits of any class are exactly the same as the class boundaries of that class.

E.g. the upper class limit of the fourth class is 40,

the upper class boundary of the fourth class is $\frac{40 + 40}{2} = \frac{80}{2} = 40$.

Similarly the lower class limit of the sixth class is 50,

the lower class boundary of the sixth class is $\frac{50 + 50}{2} = \frac{100}{2} = 50$.

The values of discrete data go up in jumps.

When discrete data is grouped into class intervals there is no need for class boundaries, since no value can lie between the limits of two classes, i.e. discrete values jump across the gap between two limits.

It is very useful to create 'artificial' boundaries for grouped discrete data. These artificial boundaries can be used during the processing of the data, e.g. plotting of cumulative frequency curves.

Similarly, ungrouped data can be considered as grouped data with one value in each interval.

Artificial boundaries can be usefully employed in the processing of ungrouped data.

The mid-value of an interval can be found using the boundaries of the interval. The mid-value obtained using class boundaries is exactly the same as that obtained using class limits.

The *size* of a class interval is the difference between the upper class boundary of the interval and the lower class boundary.

To find the size of a class interval:

Subtract the lower class boundary of the interval from the upper class boundary of the interval.

E.g. if the boundaries of an interval are 50.5 and 60.5,

then the size of the interval is given by $60.5 - 50.5 = 10$.

To decide on the size of class interval to be used when grouping data:

(*a*) find the maximum possible range of the data.

(*b*) divide this range by the number of intervals to be used.

(*c*) use class intervals of the size given by (*b*), rounded off to a suitable figure if necessary.

E.g. with a range of 60 and 12 class intervals use a class size of 5.

For a class size of 5 the difference between the boundaries of each interval should be 5.

$\uparrow$0–4 $\uparrow$ 5–9 $\uparrow$ 10–14 $\uparrow$ etc.

boundaries $\frac{1}{2}$ $4\frac{1}{2}$ $9\frac{1}{2}$ $14\frac{1}{2}$ ← difference between boundaries is 5.

With a range of 500 and 10 class intervals, use a class size of 50.

$\uparrow$0–49 $\uparrow$ 50–99 $\uparrow$ 100–149 $\uparrow$ etc.

boundaries $\frac{1}{2}$ $49\frac{1}{2}$ $99\frac{1}{2}$ $149\frac{1}{2}$ ← difference of 50

To make a *tally chart* using class intervals:

(*a*) decide on the size of each class interval or use the ones that are given.

(*b*) write the class intervals down the left hand side of the page.
(*c*) go through the data, one value at a time, as usual.
(*d*) decide which interval each value is in, as you come to it.
(*e*) put a tally against that interval in the chart.

Frequency Distribution for Grouped Data
When the totals (frequencies) for each class interval are added to the tally chart, we have the frequency distribution.
A small number of class intervals produces a 'squashed' distribution.
A large number of class intervals produces a 'stretched' distribution.
It is not advisable to produce either a 'squashed' or a 'stretched' distribution.

Example: Using class intervals of 0–9, 10–19, 20–29 etc., find the frequency distribution of the following array.

```
22 10 32  2 41  4 35 52 43 32 49 31 37 51 33
33 52 41  5 45 56 41 62 97 43 51 19  4 56 79
54 67 56 67 33 72 66 54 63 29 63 27 63 77 65
36 42 51 55 59 47 55 42  1 73 45 56 35 54 44
52 69 63 33  8 63 32  9 56 35 92 83 26 22 46
47 52  4 51 49 85 52 63 65 69 46 76 36 62 42
23 45 51 36 73 46 49 44 42 35 67 36 54 42  7
41 47  5  5 48 73 36 92 67 56 63 21 52 32 27
```

Class Interval	Tallies	Frequency
0–9	JHT JHT I	11
10–19	II	2
20–29	JHT III	8
30–39	JHT JHT JHT IIII	19
40–49	JHT JHT JHT JHT JHT I	26
50–59	JHT JHT JHT JHT IIII	24
60–69	JHT JHT JHT III	18
70–79	JHT II	7
80–89	II	2
90 and over	III	3

The frequency distribution for the same array, but using class intervals of 0–24, 25–49, 50–74, 75 and over, is as follows:

Class Interval	Tallies	Frequency
0–24	IHI IHI IHI II	17
25–49	IHI IHI IHI IHI IHI IHI IHI IHI IHI IIII	49
50–74	IHI IHI IHI IHI IHI IHI IHI IHI IHI I	46
75 and over	IHI III	8

The frequency distribution for the same array, but using class intervals of 0–4, 5–9, 10–14 etc. is as follows:

Class Interval	Tallies	Frequency
0–4	IHI	5
5–9	IHI I	6
10–14	I	1
15–19	I	1
20–24	IIII	4
25–29	IIII	4
30–34	IHI IIII	9
35–39	IHI IHI	10
40–44	IHI IHI III	13
45–49	IHI IHI III	13
50–54	IHI IHI IHI	15
55–59	IHI IIII	9
60–64	IHI IIII	9
65–69	IHI IIII	9
70–74	IIII	4
75–79	III	3
80–84	I	1
85–89	I	1
90–94	II	2
95+	I	1

The three frequency distributions appear to be different but they all show data from the same array.

The apparent differences between the frequency distributions come about because of the different number of class intervals used in each.

It is best to produce a compact distribution with about 10 classes, probably not less than 8 and not more than 15.

Grouping data does have drawbacks.

When data has been grouped some of the fine detail of the data is lost e.g. it is not possible to find the *actual* mode, median and mean of grouped data. It is only possible to *estimate* these quantities.
The disadvantages of loss of detail are normally far outweighed by the advantage of easier manageability.
If calculated with care, the estimates of the mode, median and mean of grouped data produce sufficiently accurate results for most purposes.

Cumulative Frequency Distribution for Grouped Data
To convert a grouped frequency distribution into a cumulative frequency distribution to show *less than:*

(*a*) make two new columns to the right of the frequency column as before.
(*b*) fill in the cumulative frequency column as before.
(*c*) fill in the less than column using the upper class boundaries of each interval.

Note: All of the values contained in an interval lie between the *boundaries* of the interval, so class boundaries must be used in the 'less than' column.

Example:

Class Interval	Freq.		Class Interval	Freq.	Cum. Freq.	< Less than	
1–10	2		1–10	2	2	10.5	
11–20	5		11–20	5	7	20.5	These are the
21–30	8		21–30	8	15	30.5	upper class
31–40	10	becomes	31–40	10	25	40.5	boundaries of
41–50	15	→	41–50	15	40	50.5	the intervals.
51–60	20		51–60	20	60	60.5	
61–70	18		61–70	18	78	70.5	
71–80	10		71–80	10	88	80.5	
81–90	8		81–90	8	96	90.5	
91–100	4		91–100	4	100	100.5	

To convert a grouped frequency distribution into a cumulative frequency distribution to show *more than:*

(*a*) write out the frequency distribution upside down.
(*b*) using the upside down frequency distribution, follow the same procedure as that used for less than, but fill in the more than column, using the lower class boundaries of each interval.

Class Interval	Freq.
1–10	2
11–20	5
21–30	8
31–40	10
41–50	15
51–60	20
61–70	18
71–80	10
81–90	8
91–100	4

→

Class Interval	Freq.
91–100	4
81–90	8
71–80	10
61–70	18
51–60	20
41–50	15
31–40	10
21–30	8
11–20	5
1–10	2

→

Class Interval	Freq.	Cum. Freq.	> more than
91–100	4	4	90.5
81–90	8	12	80.5
71–80	10	22	70.5
61–70	18	40	60.5
51–60	20	60	50.5
41–50	15	75	40.5
31–40	10	85	30.5
21–30	8	93	20.5
11–20	5	98	10.5
1–10	2	100	0.5

These are the lower class boundaries of the intervals.

Cumulative Frequency Distribution for Ungrouped Data

If ungrouped data is considered to be grouped data with a class size of one, then class boundaries can be used in the cumulative frequency distributions of ungrouped data.

Example: Construct a cumulative frequency distribution of the frequency distribution below.

Mark	1	2	3	4	5	6	7	8	9
Freq.	1	7	21	35	35	21	7	2	1

Considering this ungrouped data as grouped data with a class size of one, the cumulative frequency distribution to show less than is:

Mark	Freq.	CF.	Less than
1	1	1	1.5
2	7	8	2.5
3	21	29	3.5
4	35	64	4.5
5	35	99	5.5 ←
6	21	120	6.5
7	7	127	7.5
8	2	129	8.5
9	1	130	9.5

Note: There is no 'equal to' since class boundaries are being used.

————class boundaries

Similarly the cumulative frequency distribution to show more than is:

Mark	Freq.	CF.	More than
9	1	1	8.5
8	2	3	7.5
7	7	10	6.5
6	21	31	5.5
5	35	66	4.5
4	35	101	3.5
3	21	122	2.5
2	7	129	1.5
1	1	130	0.5

Note: There is no 'or equal to' since class boundaries are being used.

3.5 ←————————— class boundaries

Exercise 2c:

1. Using class intervals of 1–10, 11–20, 21–30 etc., find the frequency distribution of the following array.

41	73	31	67	50	61	33	60	23	48
22	83	38	74	47	5	55	44	54	34
37	58	94	15	90	46	37	63	20	47
21	51	46	77	25	41	50	28	52	32

2. Using class intervals of 0–9, 10–19, 20–29 etc., find the frequency distribution of the following array.

30	62	40	51	81	67	52	49	69	59
21	71	51	82	5	61	63	67	50	60
67	31	72	29	67	54	12	62	32	55
41	73	55	44	33	78	57	45	63	47
59	42	69	37	53	26	57	89	27	55
50	33	58	46	77	42	35	94	49	38

3. Construct a cumulative frequency distribution for the grouped data in question 1:

   (*a*) to show 'less than',
   (*b*) to show 'more than'.

4. Construct a cumulative frequency distribution for the grouped data in question 2:

   (*a*) to show 'less than',
   (*b*) to show 'more than'.

5. In question 1:

What is the lower class boundary of the 3rd class?
What is the upper class boundary of the 4th class?
What is the class size of each of the classes?

6. In question 2:

What is the lower class boundary of the 5th class?
What is the upper class boundary of the 7th class?
What is the class size of each of the classes?

7. Discuss with your teacher the most suitable class intervals to use in order to produce grouped frequency distributions of the:
  (*a*) heights,
  (*b*) waist sizes,
  (*c*) chest sizes,
  (*d*) hip sizes,
  (*e*) weights,
  of the pupils in your class.

Use the data you collected about your class to produce the five grouped frequency distributions listed. Would you have used the same class intervals if the data had been for a whole year group?

8. Convert each of the frequency distributions you constructed in question 7 into cumulative frequency distributions to show 'less than'.

9. Convert each of the frequency distributions you constructed in question 7 into cumulative frequency distributions to show 'more than'.

Miscellaneous Exercises 2

1. Find the frequency distribution of the following array of data.

6 5 7 4 0 1 6 2 5 4
7 6 3 6 5 4 3 5 7 5

2. Find the frequency distribution of the following array of data.

5 2 4 3 7 5 4 6 5 3 6 4 5 5 4
7 5 8 1 7 5 6 5 9 2 7 7 5 4 4
8 9 10 6 7 2 9 6 1 6 5 3 5 4 4
8 6 3 8 6 3 5 4 7 3 5 6 4 5 7

3. Find the frequency distribution of the following array of data.

```
1   2   7   1   8   8   9  10   5   5
2  10   2   5   5   4   5   2   6   4
2   4   7   7   5   4   5   2   2   5
5   5   5   4   3   3   6   3   6   2
6   8   4   2  10   2   1   5   5   6
5   3   7   4   7  12   2   3  11   8
5   3   2   4   3   5   4   3   2   3
2   2   1   7  11   8   2   7   1   7
5   7   2   3   7   4   2   7   3   0
7   6   2   9   4   2   7   2  12  10
```

4. Find the frequency distribution of the following array of data.

```
6   9   8   7   6   7  10   7   5   6
9   3   3   4   6   4   5   7   5   2
5   4   5   7   8   4   6   2   6   8
7   5   3   6   4   4   6   4   5   8
5   7   6   7   1   5   3   6   7   4
```

5. Construct a cumulative frequency distribution of the array shown in question 1.

6. Construct a cumulative frequency distribution of the array shown in question 2.

7. Construct a cumulative frequency distribution of the array shown in question 3.

8. Construct a cumulative frequency distribution of the array shown in question 4.

9. Draw up a frequency distribution, in the classes shown, for the numbers given in the table.

Frequency Distribution Table

```
2   6   1   8   7   8   1   5
2   1   4   9   6   5   7   1
7   6   8   2   9   8   5   3
4   8   6   6   7   7   1   8
9   4   3   6   2   3   8   8
```

Classes	Tally Marks	Frequency
1–3		
4–6		
7–9		

WM75

10. Draw up a frequency distribution, in the classes shown, for the numbers given in the table

4	9	9	4	5	2	5	6	3	8
1	3	4	5	7	0	5	4	6	4
8	9	9	0	4	1	6	9	3	3
1	0	7	6	3	5	0	9	1	1
5	2	7	9	4	1	8	5	6	6

Frequency Distribution Table

Classes	Tally Marks	Frequency
0–1		
2–3		
4–5		
6–7		
8–9		

WM76

11. State *one* advantage and *one* disadvantage of grouping statistical information into classes.   WM77

12. By grouping the data in classes of 0–9, 10–19, 20–29, etc. draw up a frequency distribution of the data contained in the following array.

41	60	54	92	7	63	52	69	31	50
51	81	34	84	47	61	23	37	59	85

13. In question 12:
   (a) What is the upper class limit of the 3rd class?
   (b) What is the lower class limit of the 2nd class?
   (c) What is the upper class boundary of the 5th class?
   (d) What is the lower class boundary of the 7th class?
   (e) What is the class size of the 4th class?

14. By grouping the data in classes of 0–20, 21–40, 41–50, 51–60, 61–70 and 71–100, draw up a frequency distribution of the data contained in the following array.

5	41	51	45	25	42	26	42	48	43	49	23	44	24	46
52	21	44	61	15	22	53	49	72	30	44	54	42	45	53
43	62	85	46	59	29	47	44	12	56	27	48	55	49	56
50	58	63	43	48	64	42	57	42	28	44	60	47	59	50

15. In question 14:

(*a*) What is the upper class boundary of the 2nd class?

(*b*) What is the lower class limit of the 4th class?

(*b*) What is the class size of the 2nd class?

(*d*) What is the class size of the 3rd class?

(*e*) Which class has a size of 30?

16. The table shows the first column of the frequency distribution of the lifetimes (to the nearest hour) of electric lamps.

Lifetime in hours

800–999

*1000–1199

1200–1399

What are the class boundaries of the class marked *? WM75

17. Construct a cumulative frequency distribution for the grouped data in question 9.

18. Construct a cumulative frequency distribution for the grouped data in question 10.

19. Construct a cumulative frequency distribution for the grouped data in question 12.

20. Construct a cumulative frequency distribution for the grouped data in question 14.

21. Convert the frequency distribution below into a cumulative frequency distribution.

Score	1	2	3	4	5	6	7	8	9	10
Freq.	1	4	5	15	25	32	22	8	3	2

22. Convert the frequency distribution below into a cumulative frequency distribution.

Class interval	51–55	56–60	61–65	66–70	71–75	76–80	81–85
Frequency	5	8	15	17	10	7	3

3 DIAGRAMS

Most ordinary people have difficulty in understanding tables of data.
If the results of a survey are presented in table form, most people will not fully understand the results, even if they take the time to read the tables.
People like to have things explained to them with the help of pictures.
The results of surveys can be presented in picture or pictorial form i.e. by diagrams.
If results are presented as diagrams then, provided the diagrams are simple and clear, they will be easily understood by everyone.
All diagrams should have a title and be suitably labelled.

1. Pie Chart

A *pie chart* is a circular diagram.
The circle is divided up into sectors or slices.
Each section of data is given a slice of the circle.
The size of the slice each section is given depends on the size of that section of data. The bigger the section, the bigger the slice it is entitled to.
The sizes of sections of data may appear either as frequencies or as quantities.
Note: There are 360° in a circle.

To draw a pie chart:

(a) Add up the frequencies/quantities of all of the sections of the data. This will give the total of the frequencies/quantities which are to be represented.

(b) Divide 360° by the answer to (a). This will give the number of degrees which each item/unit of data is entitled to.

(c) Multiply the frequency/quantity of each section of the data by the answer to (b). This will give the number of degrees which each section of the data is entitled to.

(d) Add up the answers to (c) to make sure that they come to 360.

(*e*) Draw a circle with *compasses*. Unless an actual radius is given, make the radius as large as possible.

(*f*) Draw a line from the outside of the circle to the centre i.e. a radius.

(*g*) Starting from the line of (*f*) use a *protractor* to divide the circle up into slices, with sizes in degrees equal to the answers from (*c*).

(*h*) Label the slices of the circle.

Note: It is important that the size of the last slice is checked against the last answer to (*c*), in case you have made a mistake in measuring the angles.

Example: On a fishing holiday, a family caught the following numbers of fish:

Fish	Perch	Roach	Carp	Trout	Salmon	Bream	Pike
Frequency	8	12	7	6	13	10	4

Show this information on a pie chart.

(*a*) 8
    12
    7
    6
    13
    10
    4
60 = Total of frequencies

(*b*)
$$60\overline{)360}$$
$$6$$
$$360$$
- - -

Therefore 1 fish = 6 degrees.

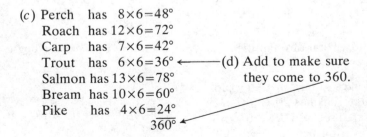

(*c*) Perch has $8\times6=48°$
Roach has $12\times6=72°$
Carp   has $7\times6=42°$
Trout has $6\times6=36°$ ←——— (d) Add to make sure
Salmon has $13\times6=78°$ they come to 360.
Bream has $10\times6=60°$
Pike   has $4\times6=24°$
                $\overline{360°}$

(*e*) Draw a circle.
(*f*) Draw a start line.
(*g*) Mark the correct angles.
(*h*) Label the slices.

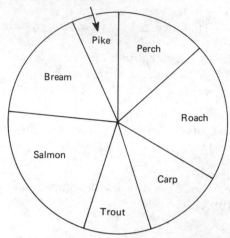

Last, so must be checked

Pike
Perch
Bream
Roach
Salmon
Carp
Trout

Pie chart to show the types of fish caught by a family on holiday

Example: An analysis of the costs involved in making a toy which sells in shops at £1.00, showed the following breakdown:

wages	35 pence
materials	20 pence
expenses	10 pence
taxation	25 pence
profit	10 pence

Show the breakdown of costs on a pie chart.

(*a*) 35
       20
       10
       25
       10
       ___
       100 = Total of quantities

(*b*)
$$100)\overline{360.0} \quad \frac{3.6}{}$$
       300
       ____
       60.0
       60.0

Therefore 1 pence = 3.6 degrees.

(*c*) Wages   has 35×3.6 = 126°
     Materials has 20×3.6 = 72°
     Expenses has 10×3.6 = 36° ←——— Add to make sure they
     Taxation has 25×3.6 = 90°   come to 360.
     Profit   has 10×3.6 = 36°
                            ____
                            360°

(*d*)

(*e*) Draw a circle.
(*f*) Draw the start line.
(*g*) Mark the correct angles.
(*h*) Label the slices.

Pie chart to show the breakdown of costs involved in making a toy

Exercise 3a:

1. A newspaper of 36 pages was found to use the following number of pages for the items listed below.

news	11	fashion	3
sport	5	editorial/letters	2
advertisements	8	finance	1
T.V./radio guide	2	features	4

Show this information on a pie chart.

2. A householder found that in his garden he had 24 rosebushes. The list below shows the numbers of each variety of rosebush in the garden.

Peace	7	Masquerade	4
Queen Elizabeth	5	Super Star	6
Iceberg	2		

Show this information on a pie chart.

3. A survey of the time given to different types of programmes by one T.V. channel during a fifteen hour period gave the following breakdown in hours.

films	4	news	1
documentaries/current affairs	2	serials/plays	3
for children	$2\frac{1}{2}$	sport	1
comedy	$1\frac{1}{2}$		

Show this information on a pie chart.

4. The survey of the types of books preferred by a year group of 120 pupils gave the following results.

thrillers	20	technical	10
romantic novels	40	detective stories	25
science fiction	15	classics	10

Show this information on a pie chart.

5. A group of 240 girls were asked about their preferences in jewellery. The results are shown below.

Type of Jewellery	Ear-rings	Pendants	Rings	Bracelets	Chokers	Necklaces	Brooches
Frequency	30	40	50	60	20	30	10

Show this information on a pie chart.

2. Line Chart

A line chart is made up of a series of lines.
Each line is separated from the other lines by a space.
Each line shows a section of the data to be represented.
The length or height of each line depends on the size of the section of the data that it is representing.
Line charts can be drawn with lines shown: (*a*) *vertically (upwards)* or
(*b*) *horizontally (across)*.
Line charts are used to show *discrete* data.
The spaces between the lines must all be the same.

Example: Draw a line chart to show the data in the frequency distribution below, which shows the colours of hair in a group of 22 people.

Colour of Hair	Blonde	Auburn	Brown	Black	Grey
Frequency	3	4	8	5	2

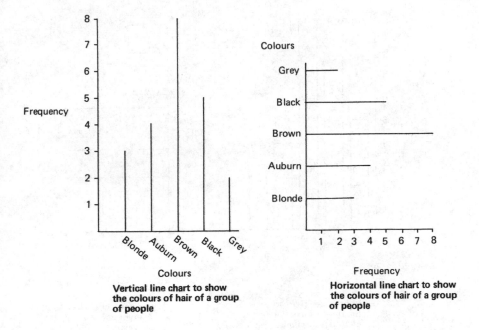

Vertical line chart to show
the colours of hair of a group
of people

Horizontal line chart to show
the colours of hair of a group
of people

3. Bar Chart

A bar chart is made up of a series of bars.
Each bar is separated from the other bars by a space.
Each bar shows a section of the data to be represented.
The length or height of each bar depends on the size of the section of data
that it is representing.
Bar charts can be drawn with bars shown: (*a*) *vertically (upwards)*
(*b*) *horizontally (across)*.
Bar charts are used to show *discrete* data.
The bars must all be the same width.
The spaces must all be the same width.
A vertical bar chart is also called a *column graph*.
(*Note*: Bars need not be the same width as spaces.)

Example: From the frequency distribution below, draw a bar chart to show
the types of meat preferred by the pupils in a class of 30.

Meat	Pork	Lamb	Beef	Chicken	Rabbit
Frequency	5	8	10	4	3

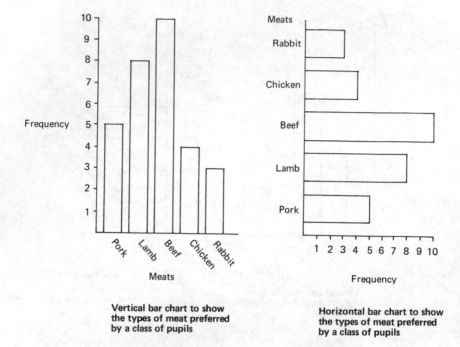

Vertical bar chart to show
the types of meat preferred
by a class of pupils

Horizontal bar chart to show
the types of meat preferred
by a class of pupils

4. Picturegram, Pictogram or Ideograph

A pictogram is a diagram made up of little drawings or symbols.
In a pictogram:

(*a*) use small sensible symbols to represent the data.
(*b*) make the symbols *simple* and *clear*.
(*c*) show the amount of data that each symbol represents i.e. a *scale*.
(*d*) make all full symbols the same size.
(*e*) make all spaces between symbols the same.
(*f*) show larger amounts of data by a greater number of symbols and not by
    larger symbols.
(*g*) draw all symbols for one section of data in one straight line.
(*h*) show data of less than the scale size by a part of a symbol, cut off across
    the direction of the line in which the symbols are running.

It is usual to draw symbols similar to the data to be represented.

Example: Horizontal lengths

Vertical lengths

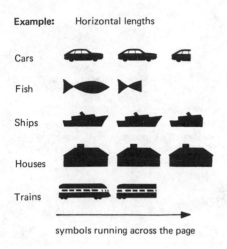

Cars

Fish

Ships

Houses

Trains

symbols running across the page

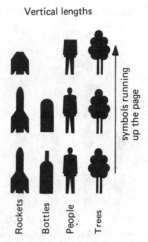

symbols running up the page

Rockets Bottles People Trees

Symbols such as cars, fish and ships should be arranged in *horizontal (across)* lines, since their lengths are in a horizontal direction.

Symbols such as people, rockets and bottles should be arranged in *vertical (upwards)* lines, since their lengths are in a vertical direction.

The reason for arranging the symbols in this way is so that the lengths of the symbols will add up to give a total length for each section of data. These total lengths show the sizes of the sections of data.

Horizontal symbols are cut off downwards.

Vertical symbols are cut off crossways.

Example: During a spot check at a school 62 pupils were found to be absent. It was later found that reasons for absence were as follows:

Reason	Flu	Coughs	Colds	Truancy	Dentist	Clinic
Frequency	16	14	8	12	7	5

Show this information on a pictogram.

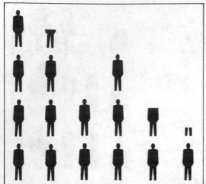

Scale 👤 = 4 pupils

Note:
On a vertical pictogram symbols are cut off crossways.

Flu Cough Cold Truancy Dentist Clinic
Vertical pictogram to show reasons for absence

Example: During one hour, the following number of cars passed a school.

Colour of Car	Red	Blue	Green	White	Yellow	Brown
Frequency	20	30	55	15	40	5

Show this information on a pictogram.

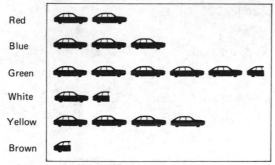

Scale 🚗 = 10 cars

Note:
On a horizontal pictogram symbols are
cut off downwards.

Horizontal pictogram to show number of cars passing a school

Exercise 3b:

1. The frequency distribution below shows how the batsmen of a local cricket team were dismissed in six games.

How Out	Bowled	Caught	Run out	LBW	Stumped	Hit Wicket
Frequency	15	20	8	11	4	2

Show this information on (*a*) a bar chart, (*b*) a line chart.

2. The following table shows the result of a survey to find the number of symphonies composed by various composers.

Composer	Beethoven	Mozart	Tchaikovsky	Shostakovich
No. of Symphonies	9	41	6	15

Composer	Sibelius	Bruckner	Brahms	Dvořák
No. of Symphonies	7	10	4	9

Show this information on (*a*) a bar chart, (*b*) a line chart.

3. The table below shows the frequencies of the different types of aircraft using an airport during one day.

Type of Aircraft	Concorde	Boeing 747	Boeing 727	BAC 1-11	Trident	Tri-Star	Light planes
Frequency	2	25	20	8	30	15	35

Show this information on (*a*) a bar chart, (*b*) a line chart.

4. The table below shows the number of horses a jockey rode to victory in each of seven years.

Year	1973	1974	1975	1976	1977	1978	1979
No. of victories	58	93	80	46	102	76	65

Show this information on (*a*) a bar chart, (*b*) a line chart.

5. The table below shows the number of trees planted by a forestry agency during one year in order to replenish stocks lost due to a fire.

Type of Tree	European Larch	Scots Pine	Douglas Fir	Spruce
No. Planted	1500	3000	10000	7000

Show this information on (*a*) a bar chart, (*b*) a line chart.

6. The noon temperatures at various places throughout the world on June 15th, 1979 are shown below.

Place	Amsterdam	Athens	Berlin	Malta	Manchester	Majorca	Paris	Moscow
Temp. °C	11	27	15	31	13	25	16	29

Show this information on (*a*) a bar chart, (*b*) a line chart.

7. Discuss suitable symbols which could be used to draw pictograms of the information in each of questions 1, 2, 3, 4, 5 and 6.

8. Collect as many statistical diagrams as you can from newspapers and magazines.
Make a wall display or make a class file of the diagrams you have collected.

5. Histogram (Block Frequency Diagram)

A histogram is made up of a series of bars.
Each bar touches the bars on either side of it.
Each bar shows a section of the data to be represented.
The bars of a histogram are drawn *vertically (upwards).*
Histograms can be used to show *continuous* data or *discrete* data.
Histograms are used mainly for continuous data.
The *area* of each bar depends on the size of the section of the data that it is representing.
The area of a bar is given by height × width.
If all of the sections of data have the same range, then the widths of all of the bars will be the same as each other.
In this case, the areas of the bars depend only on the heights of the bars, so the height of a bar can be used to represent the size of a section of data in place of area.
If the sections of data have different ranges then the widths of the bars will *not* be the same as each other.
In this case, the areas of the bars depend on both the widths and heights of the bars, so the height of each bar has to be adjusted in keeping with the width of the bar, to ensure the *correct area* is used to represent the size of each section of data.
The horizontal axis of a histogram can be marked off using:

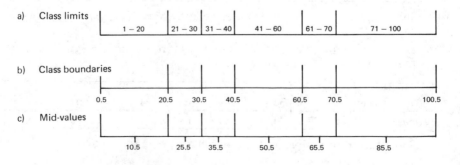

Histogram when all sections of data have the same ranges (equal class sizes).
When all of the class intervals are of equal size the vertical axis can be marked off with frequencies.
The bars are drawn using the frequencies for each interval as given by the frequency distribution.

Example: Illustrate the weight distribution below in the form of a histogram.

Weight in kg	50–51	52–53	54–55	56–57	58–59	60–61	62–63
Frequency	1	3	8	10	7	5	2

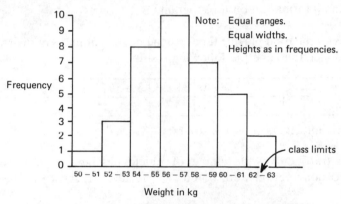

Histogram of a weight distribution

The histogram could have been drawn with class boundaries or mid-values along the horizontal axis,

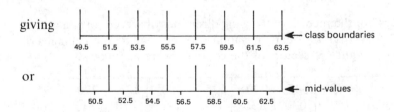

Exercise 3c:

1. A group of pupils were asked to estimate to the nearest millimetre the length of a piece of string which had a length of 4 cm. The table overleaf shows the results of their estimates.

Estimate cm	3.5	3.6	3.7	3.8	3.9	4.0	4.1	4.2	4.3	4.4	4.5
Frequency	0	5	8	14	20	27	18	9	4	2	1

Show this information on a histogram.

2. A random sample of the trees in a wood had the following circumferences (in metres).

Circumference (metres)	0–1	1–2	2–3	3–4	4–5	5–6
Frequency	4	14	26	40	12	4

Show this information on a histogram.

3. A survey of the total playing times (to the nearest minute) of the LPs in a record collection gave the following results.

Playing Time (mins)	25–27	28–30	31–33	34–36	37–39	40–42	43–45
Frequency	2	8	17	14	6	2	1

Show this information on a histogram.

4. During a traffic survey the number of vehicles passing an observer during each two hour period were noted. The results are shown in the table below.

Time Period	5–7 am	7–9 am	9–11 am	11 am–1 pm	1–3 pm	3–5 pm	5–7 pm	7–9 pm	9–11 pm
Number of vehicles	85	550	150	300	200	100	720	400	120

Show this information on a histogram.

5. An analysis of the records of the wind directions observed (over a period of 1 year) at a school weather station showed that the percentages of winds from eight 45° sectors of the compass were as follows.

Bearing (3 figure)	000–045	045–090	090–135	135–180	180–225	225–270	270–315	315–360
% winds	11	6	9	13	20	16	15	10

Show this information on a histogram.

6. A car mechanic noted the milometer readings of the cars entering his garage for M.O.T tests and made a table of his results over a period of a year.

Milometer reading	0–10,000	10,000–20,000	20,000–30,000	30,000–40,000	40,000–50,000
Frequency of cars	70	210	350	300	140

Milometer reading	50,000–60,000	60,000–70,000	70,000–80,000	80,000–90,000	90,000–100,000
Frequency of cars	60	40	20	10	5

Show this information on a histogram.

Histogram when sections of data have different ranges (unequal class sizes).

When the class intervals are not all of the same size, the vertical axis is marked off with *frequency densities* i.e. with frequencies for a given class size.

The bars for classes of this given size can be drawn using the frequencies as given by the frequency distribution.

The bars for classes which are not of the given class size must be drawn using adjusted frequencies, i.e. heights.

E.g. for a class size of half of the given class size, the frequency must be *doubled* before the bar is drawn.

For a class size twice the given class size, the frequency must be *halved* before the bar is drawn.

For a class size three times the given class size, the frequency must be *divided by three* before the bar is drawn.

Example: Draw a histogram to show the frequency distribution of the marks below.

Marks	1–20	21–30	31–40	41–50	51–55	56–60	61–70	71–100
Frequency	30	40	60	50	30	25	40	60

In the distribution above, the class intervals are not all of the same size. The frequency scale has been marked off for intervals with a class size of 10. The frequencies for class intervals with sizes other than 10 have been adjusted before being drawn, so that their *areas* give the correct overall frequencies in those intervals.

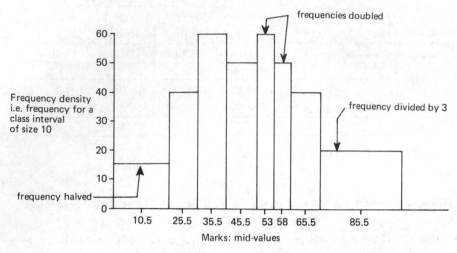

Histogram to show a mark distribution

An alternative approach to frequency density is to divide the frequency of each class interval by its class size. The frequency density is then a frequency per mark. The histogram can be drawn using these frequency densities.

Interval	Freq.	Class Size	Freq. Density Freq. ÷ Size
1–20	30	20	1.5
21–30	40	10	4
31–40	60	10	6
41–50	50	10	5
51–55	30	5	6
56–60	25	5	5
61–70	40	10	4
71–100	60	30	2

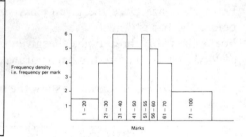

Note: The two histograms of the frequency distribution have exactly the same shape.

Exercise 3d:

1. The table below shows the frequency distribution of the shorthand speeds (words per minute) of the pupils at a secretarial college.

Speed	41–60	61–80	81–90	91–100	101–110	111–130
Frequency	40	70	70	80	50	20

Show this information on a histogram.

2. The amount of pocket money received by a class of 33 pupils is shown below.

Pocket money (pence)	50–89	90–109	110–119	120–129	130–149	150–199
Number of pupils	2	6	6	8	6	5

Show this information on a histogram.

3. An airline pilot found that cloud base (bottom of clouds) over the United Kingdom varied from day to day. For a period of six months he produced the following distribution.

Cloud base (metres)	0–300	300–1000	1000–1500	1500–3000	3000–4500
Number of days	30	45	65	25	20

Show this information on a histogram.

4. A glider pilot noted from his log book that after 200 solo flights the duration of his flights (in minutes) were as follows.

Duration (mins.)	0–5	5–10	10–20	20–30	30–60	60–120
Frequency	40	30	50	32	24	24

Show this information on a histogram.

5. An analysis of a sample taken from a beach showed the following composition by weight in g.

Description (Atterberg scale)	Clay	Silt	Fine Sand	Coarse Sand	Gravel
Size of particles (diameter mm)	under 0.002	0.002–0.02	0.02–0.2	0.2–2.0	2.0–20
Weight g	2	8	200	1200	1000

Show this information on a histogram.

6. An analysis of the records of the wind strengths in knots at a school weather station over a period of 1 year gave the following table.

Wind strength (knots)	less than 1	1–10	11–21	22–33	34–71
Percentage	8	40	36	12	4

Show this information on a histogram.

7. The Beaufort scale equivalents for the above question would be

Beaufort wind strength	0	1–3	4–5	6–7	8–12
Percentage	8	40	36	12	4

Would the histogram of this data have a different shape to that of question 6? If so why?

6. Frequency Polygon

A frequency polygon is often used in place of a histogram.
A frequency polygon shows the general shape of the frequency distribution.
If the class intervals of a distribution are all of *equal class size*, a frequency polygon encloses the *same area* as the histogram that it is replacing.
Frequency polygons are useful for comparing distributions of similar kinds of data.
To draw a frequency polygon:

(*a*) draw a histogram of the data.
(*b*) mark the middle point of the top of each bar.
(*c*) at either end of the histogram mark the horizontal (across) axis where the mid-value of an extra (i.e. imaginary) bar would be situated. Each of the two imaginary bars must be taken to have the same class size as the bar adjacent i.e. next to it.
(*d*) join the marked points from (*c*) and (*b*) with straight lines.

Example: Draw a frequency polygon of the marks obtained by a group of pupils as detailed by the distribution below.

Mark	1–10	11–20	21–30	31–40	41–50	51–60	61–70	71–80	81–90	91–100
Frequency	2	4	5	7	10	8	7	4	3	1

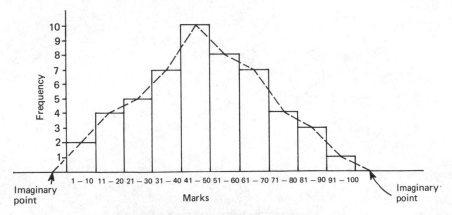

The dotted line is the Frequency polygon of the distribution of marks.

A frequency polygon can be drawn without drawing a histogram.
To draw a frequency polygon in this way:

    (*a*) plot points where the middle points of the tops of the bars of a histogram would be i.e. mid-values against frequencies.

    (*b*) plot two imaginary points as before.

    (*c*) join the plotted points from (*a*) and (*b*) with straight lines.

This method can be used when two distributions are to be compared.

Example: The frequency distributions below, show the marks obtained by the pupils in the first form of a school in two English tests. The first test was given at the beginning of the year, the second test was given at the end of the year.

Test 1.

Mark	1–10	11–20	21–30	31–40	41–50	51–60	61–70	71–80	81–90	91–100
Frequency	5	20	25	50	40	20	15	10	10	5

Test 2.

Mark	1–10	11–20	21–30	31–40	41–50	51–60	61–70	71–80	81–90	91–100
Frequency	5	5	10	20	35	50	30	20	15	10

On the same diagram draw a frequency polygon for each set of marks.

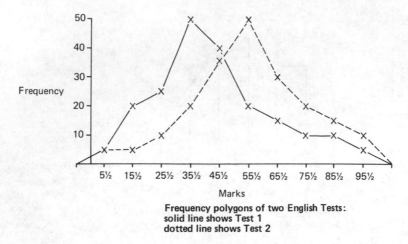

Marks

**Frequency polygons of two English Tests:
solid line shows Test 1
dotted line shows Test 2**

The frequency polygons above, show that overall the first form have improved in English during the year. Discuss this statement with your teacher.

Exercise 3e:

1. Construct frequency polygons of the histograms of Exercise 3c.

2. Construct frequency polygons of the histograms of Exercise 3d.

7. Trend Graphs, Time Series Graph or Historigram

If a statistical quantity changes, i.e. goes up and down as time goes by, it is usual to show the changes on a *trend* graph. A trend graph, as its name implies, shows how the quantity being represented is trending, i.e. moving upwards or downwards, as time goes by.

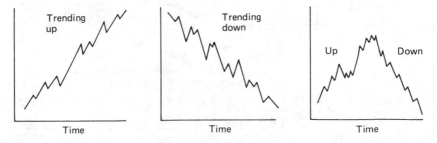

Time can be measured in many ways: seconds, minutes, hours, days, weeks, months, years.

Examples of Trends:
Rainfall usually goes up during the winter months and goes down during the summer months.
Ice cream sales usually go up during the summer months and go down during the winter months.
The traffic on a road usually goes up as rush hour approaches and goes down after the rush hour.
To draw a trend graph:

(*a*) draw two axes on graph paper, one vertical and one horizontal.
(*b*) mark off the horizontal (across) axis in time units.
(*c*) mark off the vertical (up) axis in the units of the quantity which is changing with time.
(*d*) plot each pair of values in the list of data against each other, marking the position of each pair of values with a cross on the graph.
(*e*) join up the crosses on the graph with *straight lines*.

Example: The number of Christmas cards delivered at a block of four flats during December is shown below. Show this information on a trend graph.

Date Dec.	5	6	7	8	9	10	11	12	13	14	15	16	17	18	19	20	21	22	23	24
No. Cards	2	6	8	9	5	6	8	10	7	10	12	8	11	11	13	15	13	14	20	15

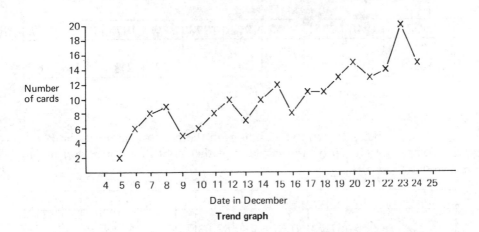

Trend graph

Example: The amount of money shown on a man's bank balance at the end of each month is shown below. Show this information on a trend graph.

Month End	Jan	Feb	Mar	Apr	May	Jun	Jul	Aug	Sep	Oct	Nov	Dec
Balance £'s.	50	40	70	100	80	150	200	10	50	80	120	20

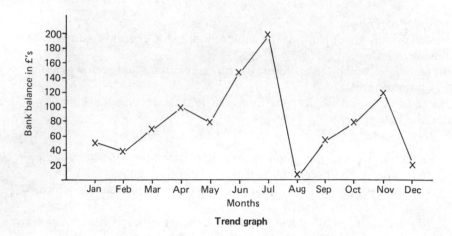

Trend graph

The above graph shows two trends upwards, with dramatic drops in August and December. Why do you think these drops occur?

Exercise 3f:

1. The following table shows the decline in the purchasing power of the pound in various years since 1946 (taking 1946 as the base year).

Year	1946	1951	1967	1972	1973	1974	1975	1976	1977
Purchasing Power (pence)	100	76	47	$34\frac{1}{2}$	$31\frac{1}{2}$	27	$21\frac{1}{2}$	19	16

Show this information on a trend graph.

2. The table below shows the variations in the price of shares in the PDU Computing Group on the last day of the month over a period of 18 months.

Month	J	F	M	A	M	J	J	A	S	O	N	D	J	F	M	A	M	J
Share Price (pence)	52	57	72	60	67	63	80	82	83	70	70	78	69	84	95	93	88	92

Show this information on a trend graph.

3. The table below shows the heights of the high tides at the resort of High Water over a period of 34 days.

Date	1	2	3	4	5	6	7	8	9	10	11	12
High Tide (metres)	5.3	5.5	5.9	6.2	6.5	6.7	6.7	6.6	6.6	6.4	6.2	5.9

Date	13	14	15	16	17	18	19	20	21	22	23	24
High Tide (metres)	5.5	5.2	4.9	4.9	5.1	5.4	5.7	6.0	6.4	6.8	7.0	7.2

Date	25	26	27	28	29	30	31	1	2	3
High Tide (metres)	7.1	7.0	6.7	6.3	5.8	5.4	5.2	5.3	5.5	5.9

Show this information on a trend graph.

4. The maximum temperatures recorded each day for a fortnight at a school weather station are shown below.

Day	M	T	W	T	F	Sa	Su
Max. Temp °C	18	22	21	18	18	20	19

Day	M	T	W	T	F	Sa	Su
Max. Temp °C	27	17	25	16	17	21	20

Show this information on a trend graph.

5. The hours of sunshine per day at the station over the same fortnight are shown below.

Day	M	T	W	T	F	Sa	Su
Sun (Hours)	12.6	11.4	8.2	7.3	13.3	10.5	13.4

Day	M	T	W	T	F	Sa	Su
Sun (Hours)	13.7	12.2	13.1	14.2	14.0	10.3	15.1

Show this information on a trend graph.

6. The grass pollen count (of great interest to hay fever sufferers) is given by the G.P.O. each day during the pollen period. The count for Haytown over a period of 35 days is detailed in the table below.

Day Number	1	2	3	4	5	6	7
Pollen Count	5	8	15	10	25	80	120

Day Number	8	9	10	11	12	13	14
Pollen Count	70	95	115	175	190	160	140

Day Number	15	16	17	18	19	20	21
Pollen Count	175	250	290	300	280	270	300

Day Number	22	23	24	25	26	27	28
Pollen Count	200	180	150	160	180	120	140

Day Number	29	30	31	32	33	34	35
Pollen Count	80	50	60	30	30	10	20

Show this information on a trend graph.

8. Cumulative Frequency Curve or Ogive

A *cumulative frequency curve* is a graph showing the cumulative frequencies of a set of data.
The points on the graph are connected by a *smooth curve*.
To draw an ogive of a set of data:

(*a*) construct a cumulative frequency distribution of the data using *class boundaries*.
(*b*) draw two axes on graph paper.
(*c*) mark off the horizontal (across) axis in the units of either the 'less than' column *or* the 'more than' column.
(*d*) mark off the vertical axis (up) as cumulative frequencies.
(*e*) plot each of the *class boundaries* of the last column against its cumulative frequency. Mark the position of each *pair* of plotted values with a cross.
(*f*) join up the crosses on the graph with a smooth curve.

Note: Take care to plot the cumulative frequencies vertically.

Example: Draw an ogive of the cumulative frequency distribution below.

Class	Freq.	CF	Less than
0–9	1	1	9.5
10–19	3	4	19.5
20–29	5	9	29.5
30–39	10	19	39.5
40–49	14	33	49.5
50–59	13	46	59.5
60–69	10	56	69.5
70–79	5	61	79.5
80–89	3	64	89.5
90–99	1	65	99.5

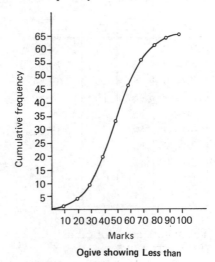

Ogive showing **Less than**

The points are plotted as:
(9.5, 1) (19.5, 4) (29.5, 9)
(39.5, 19) (49.5, 33) (59.5, 46)
(69.5, 56) (79.5, 61) (89.5, 64)
(99.5, 65).
The origin (0, 0) is taken as an additional point.

Example: Draw an ogive of the cumulative frequency distribution below.

Mark	Freq.	CF	Less than
1	1	1	1.5
2	7	8	2.5
3	21	29	3.5
4	35	64	4.5
5	35	99	5.5
6	21	120	6.5
7	7	127	7.5
8	2	129	8.5
9	1	130	9.5

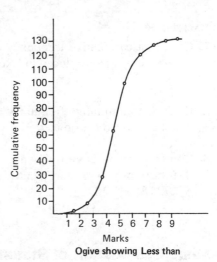

Ogive showing **Less than**

The points are plotted as:
(1.5, 1) (2.5, 8) (3.5, 29)
(4.5, 64) (5.5, 99) (6.5, 120)
(7.5, 127) (8.5, 129) (9.5, 130).
The origin (0, 0) is taken as an additional point.

Example: Draw an ogive of the cumulative frequency distribution below.

Class	Freq.	CF	More than
90–99	1	1	89.5
80–89	3	4	79.5
70–79	5	9	69.5
60–69	10	19	59.5
50–59	13	32	49.5
40–49	14	46	39.5
30–39	10	56	29.5
20–29	5	61	19.5
10–19	3	64	9.5
0–9	1	65	−0.5

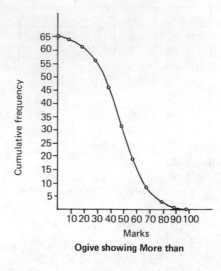

Ogive showing More than

The points are plotted as:
(89.5, 1) (79.5, 4) (69.5, 9)
(59.5, 19) (49.5, 32) (39.5, 46)
(29.5, 56) (19.5, 61) (9.5, 64)
(0, 65).

Note: The final point is plotted as (0, 65) and the point (99.5, 0) is taken as an additional point.

Exercise 3g:

1. Construct cumulative frequency curves from the cumulative frequency distributions you produced in Exercises 2a questions 8, 9, 10, 11, 12, 13, 14 and 15.

2. Construct cumulative frequency curves from the cumulative frequency distributions you produced in Exercises 2c questions 3, 4.

3. Construct cumulative frequency curves from the cumulative frequency distributions you produced in Miscellaneous exercises 2 questions 5, 6, 7, 8, 17, 18, 19, 20, 21, 22.

Misrepresentation of Statistics

Statisticians are highly trained workers and the statistics they produce are worked out with care. Their statistics give as accurate a picture of a given situation as it is possible to obtain.

The statistics produced by statisticians are used by many people, some of whom are less than honest. These less than honest people tend to present statistics in a light favourable to themselves. They misrepresent statistics to distort the true picture of a situation to suit themselves.

When presented with information a questioning attitude should be adopted.

(*a*) Who says so?
(*b*) What does he have to gain by making me believe the information?
(*c*) Is he trying to mislead me?

Statistics are frequently misrepresented in advertisements. The following questions should be asked about how the data was collected.

(*a*) Was the sample large enough?
(*b*) Was the sample random?
(*c*) Was the sample likely to be biased in any way?

E.g. in a survey it was found that 9 out of 10 people prefer Whizzo.

(*a*) How many people were involved in the survey, 10 or 10,000?
(*b*) Were the people questioned just after they took a packet of Whizzo down from the shelf in the supermarket?
(*c*) Were the people questioned offered a free supply of Whizzo and thought they must answer the question 'correctly'?

Diagrams are often used to misrepresent data. The following questions should be asked when considering diagrams.

(*a*) Is data being presented on a satisfactory type of diagram?
e.g. Is a bar chart used when a histogram should be used?
(*b*) Is the diagram chosen to represent the data as it should be?
e.g. Is a scale on any axis missing?
Do scales on the axes start at zero or has the origin been altered to gain a dramatic effect?
Have the scales on the axes been broken at any point?
Have the scales on the axes been expanded or decreased anywhere?
On a pictogram have bigger symbols been used to show more data, thus gaining dramatic effect, instead of using more symbols of the same size?
(*c*) Can dramatic changes in a graph be accounted for by seasonal variations, as in the sales of ice-cream during the summer?

Miscellaneous Exercises 3

1. A survey of the type of holiday taken last year by the 30 pupils in form 4x gave the following information:

holidays abroad   14 pupils
seaside holiday in Gt. Britain   6 pupils
other holiday in Gt. Britain   8 pupils
no holiday   2 pupils

Represent the above information as a pie chart, carefully showing any calculations you do. Mark on your diagram the sizes of the angles you have calculated.   Y77

2. In a survey it was found that the total domestic expenditure of all the households questioned was as shown in the table below:

	£1000's
Food	60
Housing	20
Clothing	15
Tobacco and Drink	17
Entertainment	8

Represent this information by means of:
 (i) A pie chart.
 (ii) A bar chart.   EM78

3. The table shows the number of hours per week that a pupil spends on school subjects. Present this information in the form of a pie chart.

English   3
French   3
mathematics   4
physics $2\frac{1}{2}$
chemistry $2\frac{1}{2}$
other subjects   9   WM75

4. A factory worker spends 55% of his waking life at work, and the rest of the time elsewhere. Present this information in the form of a pie chart.
                                                                      WM76

5. (*a*) Illustrate the figures in Table A below by drawing an accurate pie-chart, using a circle of radius 6 cm. Label your chart clearly.

Table A
The way in which John spent 24 hours

Activity	Number of hours spent on it
Sleep	9
Work	7
Play	3
Meals	$2\frac{1}{2}$
Other activities	$2\frac{1}{2}$

(*b*) Table B below shows the number of males and females attending a new leisure centre in three recent weeks.

Table B

| Week | Number of | |
	Males	Females
1st	2250	1500
2nd	5500	5000
3rd	8000	6500

Illustrate this information by drawing pictograms, giving a suitable title to the completed illustration.

(*c*) Table C below shows the frequency distribution of the number of days pupils in a class were absent during a certain term.

Table C

Number of days absent	2 or less	3	4	5	6 or more
Frequency of pupils	6	3	5	6	10

Construct a histogram which illustrates this distribution. Explain how to deal with the frequencies given for the number of days absent described as '2 or less' and the one described as '6 or more' and comment on any difficulty that arises in interpreting these.   M77

6. A company made an overall profit of £2,000,000 during the four years from 1973 to 1976. In 1973 the company made a loss of £400,000 but in the years 1974 and 1975 it made profits of £600,000 and £800,000 respectively.

(*a*) Find the profit made in 1976.

(*b*) Using graph paper and a scale of 1 cm to £200,000, draw a bar chart with bar width of 1 cm to show the annual profit or loss of the company for each of the four years.   M77

7. (*a*) Table A below shows the countries to which the passengers in 120 cars leaving a cross-channel ferry boat were going to spend their holidays.

Table A

Country	Number of cars
France	45
Germany	18
Holland	10
Spain	27
Italy	20

Illustrate this information by drawing an accurate pie chart, using a circle of radius 6 cm. Label your chart clearly.

(*b*) Table B below shows the distances, in kilometres, travelled by the same 120 cars.

Table B

Distance travelled (km)	Number of cars
less than 800	20
800–1200	10
1200–1600	15
1600–2000	12
2000–2400	10
2400–2800	25
2800–3200	12
more than 3200	16

Construct a histogram which illustrates this information. Show any calculations which need to be carried out before the histogram can be drawn. Comment on any difficulties you might have in interpreting the way the ranges of distances have been recorded in the table.   M78

8. (*a*) In a certain borough each £1 of the General Rate collected is spent as follows:

Education	47.5p
Public Health	21.0p
Police and Fire Service	12.5p
Housing	10.0p

The remainder is spent on the provision of other services.
Illustrate this information by means of a bar chart.

(*b*) Each member of a group of 24 girls was asked to name her favourite type of T.V. programme from a given list of five. The results were represented on a pie chart and the angles of the sectors were as follows:

comedy	75°
drama	120°
current affairs	45°
crime series	30°

The remaining sector represented sport. How many girls named sport as their favourite type of programme? L78

9. Use the information given in table A and B to answer (*a*) and (*b*) below.

Table A
The main methods used by 240 children for getting to school

Method of getting to school	Number of children
Bus	88
Cycle	64
Walking	40
Car	32
Train	16

Table B
The frequency of the times taken by 240 children to get to school from their homes

Time, in minutes	Less than 2	2–5	5–10	10–15	15–20	20–25	25–30	More than 30
Frequency	10	21	40	68	50	25	14	12

(*a*) Draw an accurate pie-chart, using a circle of radius 6 cm, to illustrate the figures in table A. Label your chart clearly.

(*b*) Draw a histogram to illustrate the figures in table B. Show any calculations which need to be carried out before the histogram can be drawn. Comment on any difficulties that you might have in interpreting the way that the time intervals have been recorded here.

M76

10. An investigation was carried out into the reasons for absence or for late arrival at a school on a particular Monday morning. The results of the investigation were as follows:

missed the bus	25%
sickness	38%
overslept	11%
had a cold	21%
bicycle had puncture	5%

(*a*) On the graph paper provided, draw a histogram to illustrate these statistics. Use a scale of 2 cm to represent 5% on the vertical axis.
(*b*) If the information had been represented on a pie chart, calculate the angle at the centre which would have been required for the sector representing 'sickness'.
(*c*) If 1200 pupils attended this school and, on that particular morning, 20% were absent or arrived late, calculate how many pupils missed the bus.

Y77

11. For the data given below, draw the histogram on the diagram given.

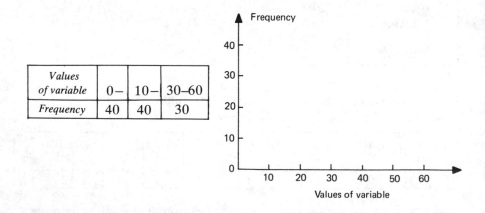

Values of variable	0–	10–	30–60
Frequency	40	40	30

WM77

12. From the data given below, complete the histogram on the diagram on the right.

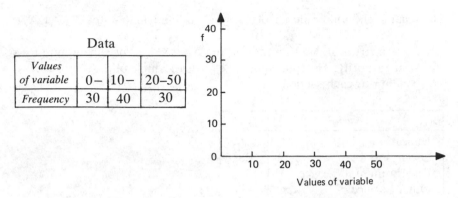

Data

Values of variable	0–	10–	20–50
Frequency	30	40	30

WM75

13. The frequency distribution of the number of matches in each of 41 boxes is as shown below:

Frequency	4	2	4	7	8	6	6	4
No. of matches	31–34	35–36	37–38	39	40	41	42–43	44–45

Plot a histogram to show this information.   EM78

Questions 14–16 concern the following histogram, which represents a stock of nails held by a handyman. Column X shows a total of 200 nails for the class 0–5 cm.

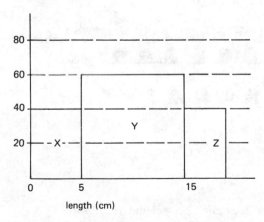

14. How many nails are represented by column Y?

15. If column Z represents 120 nails, what is the upper boundary (end point) of the class 15 cm upwards?

16. What is the total number of nails in the handyman's stock?   EA78

17. A pie chart is drawn to represent the five sections of expenditure by a local authority. The following table gives the angle of the sector corresponding to each section.

Section	Angle
Education	216°
Social Services	28°
Environmental Services	36°
Other Services	30°
Inflation	

(*a*) Calculate the angle of the sector corresponding to Inflation.
(*b*) The total expenditure by the local authority was £94.5 million. Calculate the amount spent on the Social Services.
(*c*) Calculate the expenditure on Education as a percentage of the total expenditure.
(*d*) Using a radius of 5 cm draw the pie chart, labelling the sectors clearly.   WM78

18. The following pictogram, which has been left uncompleted, represents the number of civil servants employed in various Ministries.

	Number of civil servants
Department of Health & Social Security	90 000
Inland Revenue	75 000
Home Office	
Environment	28 000

(*a*) How many civil servants does each bowler hat represent?
(*b*) Use this scale to find the number of civil servants employed in the Home Office.

(*c*) If 28,000 civil servants are employed in the Department of the Environment, complete the diagram above to represent this information. Y78

19.

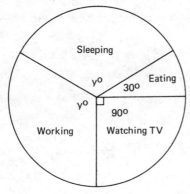

The pie chart shown, not drawn to scale, represents a day in the life of a Middlesex schoolboy.

(*a*) Find the time, in minutes, represented by an angle of 1°.

(*b*) Calculate how many hours per day he spent watching T.V.

(*c*) Calculate the time in hours spent in working each day. M78

20. The cost of a new school textbook is £3.20. This cost is broken down into 4 parts as shown in the pie diagram below.

Calculate, to the nearest penny,

(*a*) the writer's fee,

(*b*) the cost of publishing.

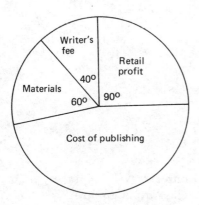

Y78

21. The diagram shows a pie chart drawn to illustrate the numbers of people in four groups A, B, C, D.

(*a*) Calculate the value of x.

(*b*) If there are 49 people in group B how many are there in group A?

(*c*) If group B is now subdivided into two new groups whose numbers are in the ratio of 2:3, calculate the angles for the new groups.

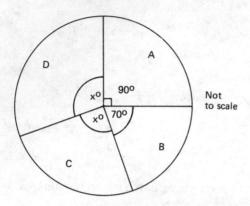

Not to scale

WM76

22. A class of 32 children is divided into four groups A, B, C, D. The diagram below shows a column graph drawn to illustrate the numbers in each group.

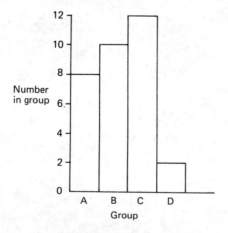

(*a*) How many children are there in Group D?

(*b*) What percentage of children in the class are in group A?   Y77

23.

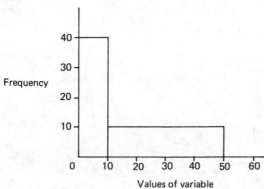

Use the histogram given here to complete the frequency table below.

Values of variable	0–	10–50
Frequency	40	

WM76

24.

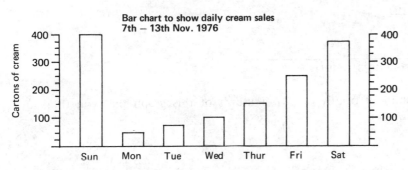

(a) (i) Copy out the table shown below and complete it from the bar chart.

Day	Sun.	Mon.	Tues.	Wed.	Thur.	Fri.	Sat.
No. of Cartons							

(ii) Explain the shape of the chart.

(iii) Why is such a chart useful to the dairy?

(*b*) Illustrate the following data by means of a pie-chart. Use a circle of radius 6 cm and show your calculations.

The Results of a Pie Popularity Poll

Type of Pie	No. of Votes
Gooseberry	700
Blackcurrant	200
Apple	500
Blackberry	1000
Cherry	1200

NI77

25. The following diagram represents a sales graph for a brand of petrol called 'Zoom'.

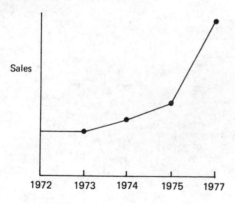

Comment briefly on *two* features of this graph that you think are incorrect.

Y78

26. The values of sales of a dog-food called Woofmeat were as follows:

Year	Sales (£)
1975	1000
1976	2000
1977	3000

In an advertising campaign, the makers of Woofmeat produced the following diagram:

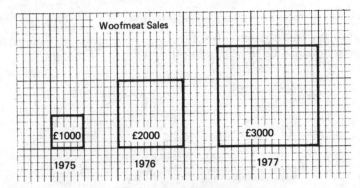

Explain how the diagram gives an impression of a much larger increase in sales than actually took place. EA78

27. Give *one* reason why it is sometimes an advantage to represent statistical data in pictorial form.

28. An automatic washing machine has a programme in which the following takes place:

filling with water   8 minutes
washing   15 minutes
emptying water   8 minutes
rinsing   10 minutes
spinning   4 minutes

(*a*) Draw a pie-chart to show this information.
(*b*) Give *one* other way in which this information could have been displayed. NW78

29. (*a*) The frequency distribution of the shoe sizes of a class of 30 girls is represented by the table below.

Shoe size	$3\frac{1}{2}$	4	$4\frac{1}{2}$	5	$5\frac{1}{2}$	6
No. of girls	3	5	10	7	4	1

(i) Draw a histogram for this data.
(ii) What is the mode of the distribution?

(*b*) Each member of the fifth year boys was asked to choose a favourite sport from the list below. Represent their replies in a pie chart.

soccer   46
rugby   33
cricket   25
hockey   16

(*c*) Statistics may sometimes be misrepresented.

For each of the three graphs shown below say why the statement underneath may not be an accurate interpretation of the graph.

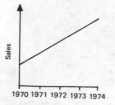 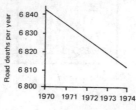 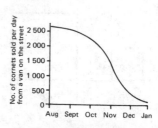

(i) The graph shows that sales have more than doubled between 1970 and 1974.

(ii) This graph proves that there has been an enormous reduction in deaths on the road over four years.

(iii) The graph shows that the standard of the ice cream is declining as the van is selling a lot less of the product.

NW75

30. (*a*) A teacher stated that the two most used letters in the alphabet were *P* and *E*. As an example he asked the class to consider: 'If Peter Piper picked a peck of pickled pepper where is the peck of pickled pepper that Peter Piper picked?' The frequency of the various letters is *E*–19, *P*–18, *I*–8, *R*–7, *C*–6, *K*–6, *T*–4, *D*–4, *H*–3, *F*–3, *L*–2, *A*–2, *O*–2, *T*–1, *S*–1, *W*–1, and that shows that *E* and *P* are the two most frequently used letters. Make *three* comments on the statement.

(*b*) A manufacturing company produced the chart shown to impress people with their sales expansion over the last 30 years. The sales manager had joined the firm in 1960 and you can see from the graph how the line is steepest from 1960 onwards.

(i) What comments would you make about the scales?

(ii) Draw the graph with correct scales, choosing suitable ones for the graph paper. What do you think your graph shows?

(iii) In 1975, of the 12 million sales, 8 million were on the home market, 3 million to the African market and 1 million to the Australian market. Illustrate this with a pie chart.

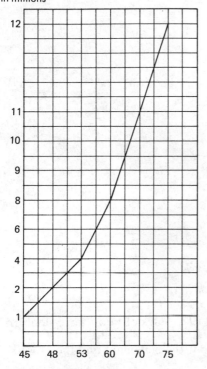

Gross sales
in millions

NW77

31. (*a*) On a long bus journey a passenger kept a check on the distance
travelled on various types of roads and his results were as follows:

Type of road	Kilometres travelled
Motorway	75
Dual Carriageway	50
3 lane A road	20
2 lane A road	25
B road	10

Represent this information on graph paper in these following three ways:
  (i) a pie chart
 (ii) a bar chart
(iii) a pictograph.
Use approximately half a side of graph paper for each diagram.
State clearly in each case the scale you are using and show clearly any
necessary calculations.

(For the pie chart use a circle of radius 5 cm and for the pictograph let

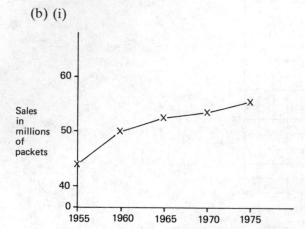

= 20 km.)

(b) (i)

Sales in millions of packets

Sales figures of the washing powder 'FIZ'

'In the 20 year period from 1955 to 1975 sales of "Fiz" have doubled.'

Make a total of *two* comments about the statement and the graph.

(ii)

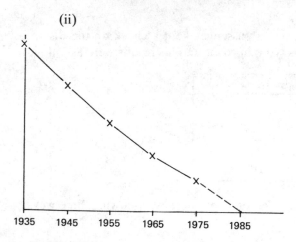

Number of infant deaths at birth per 1000 births

'In 10 years' time it can be seen that no infants will die at birth.'

Make a total of *two* comments about the statement and the graph.

NW78

4 AVERAGES

We saw earlier that when collecting data it is often useful to choose a sample to represent a population.

Similarly, when data has been tabulated, it is useful to choose a single value or number to represent a frequency distribution.

The value chosen to represent a distribution is called the average value of the distribution.

The data in any distribution is dispersed, i.e. spread out, over a range of values.

The average value is usually a value close to the centre of this range.

Since the average is usually a value close to the centre of the distribution it is a measure of central location, or a measure of central tendency of the distribution.

A statement such as

'The average man is 5 ft. 8 ins tall',

does not mean that every man is 5 ft. 8 ins, but that the heights of men are spread out around the height 5 ft. 8 ins.

Other examples of statements indicating central location might include:

The average weekly wage packet in England is £85.

The average number of matches per box is 52.

The average price of houses in England is £18,000.

The average man takes size 7 in shoes.

There are three main ways of working out the average of a frequency distribution.

Each one of the three ways produces a different type of average.

The three types of averages are called:

(a) mode   (b) median   (c) arithmetic mean.

Each one of the three types of average has its own:

(a) uses   (b) advantages   (c) disadvantages.

Mode

The *mode* of a distribution is the *value* (thing, item) in the distribution which turns up the *most*.

This value is also called the *modal* value.

The mode is very useful when the items in a distribution do not have numerical values, e.g. colours, letters etc.

To find the *mode* of a set of data:

(*a*) construct the frequency distribution of the data.

(*b*) find the value with the largest frequency, i.e. the value which turned up the most.

(*c*) write down the answer to (*b*) as the mode.

Example: Find the mode of the following array.

```
 8  3  1  3  5  6  2  6  0  5
 4  5  3  6  6  4  3  8  6  5
 5  5  7  5  2  4  9  5  9 10
10  8  8  9  4  7  8  9  7  4
```

The frequency distribution is:

Number	Tallies	Frequency
0	I	1
1	I	1
2	II	2
3	IIII	4
4	HHI	5
5	HHI III	8
6	HHI	5
7	III	3
8	HHI	5
9	IIII	4
10	II	2

Mode ———→ 5 ... 8 ←——— Most

The number which turns up the most is 5.

The mode of the array is 5.

The modal value of the array is 5.

A distribution with one mode is called a *uni-modal* distribution.

It is possible for a distribution to have more than one mode.

A frequency distribution may show signs of separation into two recognisable sections, e.g. the heights of a group of adults or the number of cars per hour passing through a set of traffic lights during a day.

The histogram or frequency polygon of a distribution in which separation occurs has a double humped appearance.

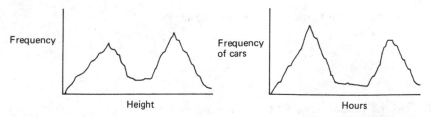

Separation due to the different height ranges of men and women.

Separation due to the heavy traffic occurring at rush hours.

Each section of the distribution is considered to have a mode.

The mode of a section of a distribution is the value in that section with the largest frequency.

Double humped distributions have two modes, i.e. one for each section.

A distribution with two modes is called a *bi-modal* distribution.

A distribution which does not show separation may be bi-modal.

This occurs when two values in the distribution both turn up equally most often.

Example: Consider this sentence and then find which is the modal vowel contained in it.

The frequency distribution of the vowels in the sentence is:

Vowels	Tallies	Frequency
a	III	3
e	IIII III	8
i	IIII III	8
o	IIII	4
u		0

Mode ⟶ e — Equal most
Mode ⟶ i — Equal most

Two vowels turn up equal most so the distribution is bi-modal.

The two modes are e and i.

Exercise 4a:

1. Find the modal value in each of the following series of numbers.

```
(a) 1  3  5  3  2  4  3  6  6  7
(b) 1  5  2  3  5  7  5  5  9  8
(c) 1  3  5  2  5  3  3  5  8  5
(d) 1  6  5  4  7  4  6  4  4  8  6
(e) 2  0  9  3  2  2  7  5  3
(f) 9  6  3  5  2  5  5  4  8  5  4  4  5
(g) 3  1  3  2  2  3  7  4  3  9  3  7  3  6
(h) 4  5  4  1  0  4  3  4  2  4  0  5
```

2. Find the modal value of the following array.

```
1 6 4 5 6 5 4 6 4 7
8 4 9 4 6 2 3 5 7 4
3 8 3 4 1 7 9 3 7 5
```

3. Find the modal value of the following array.

```
5 7 4 6 1 7 6 8 4 6
1 9 5 9 5 6 1 5 4 7
7 5 4 5 6 5 5 6 5 8
1 6 6 9 5 5 6 8 7 5
```

4. Find the modal value of the following array.

```
9 5 3 1 2 4 1 8 2 3 5
2 3 6 4 3 5 3 3 4 9 5
3 2 3 8 2 4 6 4 3 3 4
```

5. The dental records of 45 pupils showed that they each had a number of fillings. The numbers of fillings each had is given in the array below.

```
1 5 3 2 4 6 5 4 4 6 4 7 4 4 3
5 7 8 3 6 3 2 1 5 3 3 4 4 7 2
4 6 4 5 4 4 4 5 4 2 4 5 3 3 4
```

Find the modal number of fillings per pupil.

6. On a biology field trip 75 birds' nests were examined and the number of eggs in each was noted. The results are shown in the array below.

```
3 4 3 2 4 2 1 4 3 5 4 0 0 4 4
6 6 4 5 4 5 4 3 5 4 4 3 5 2 3
3 4 0 3 4 4 2 5 2 1 4 3 3 4 5
7 6 4 0 5 2 4 1 3 4 3 5 2 4 4
6 3 6 4 0 3 4 3 2 4 3 3 4 4 5
```

Find the modal number of eggs per nest.

7. A motorist made a note of the amount of petrol (in gallons) he used each week over a period of a year. The array below is a copy of his notes.

```
5  5  7 8 4 5 3 8 4  5 8  7  4
9  1 11 9 6 3 6 7 7 15 6 10  6
8 13  7 2 7 6 3 5 8  7 5  4 10
7 10  6 9 7 8 7 4 6  6 7  7  5
```

Find the modal number of gallons used per week.

8. The noon temperatures in °C in Rome over a period of 30 days are shown below.

27 31 29 27 26 25 24 26 25 26 21 27 20 28 25
22 28 26 23 23 27 22 30 27 27 31 31 27 20 28

Find the modal noon temperature.

9. A frequency distribution of the number of pages in a set of paperbacks written by the same author is shown below.

No. of pages	248	249	250	251	252	253	254	255	256	257	258	259	260
No. of books	1	2	4	8	3	3	3	2	0	2	1	1	1

What is the modal number of pages per book written by this author.

10. A knitting pattern for the back of a ladies' jumper gave the following instructions.

No. of stitches	60	58	56	54	52	49	47	42	37	32	27	23	19
No. of rows	34	1	1	2	2	2	29	2	2	2	1	1	1

Find the modal number of stitches per row.

11. A rock climbing guide book listed the heights of the various climbs on a rock outcrop to the nearest 2 m. A frequency distribution of the heights of the first one hundred climbs on the outcrop is given below.

Height (m)	6	8	10	12	14	16	18
Frequency	30	26	19	9	9	5	2

Find the modal height of the climbs.

12. A top class golfer played 80 rounds in 20 competitions. The frequency distribution below shows his scores.

Score	66	67	68	69	70	71	72	73	74	75	76	77	78
Frequency	1	2	3	8	20	6	10	10	8	5	3	2	2

Find the golfer's modal score.

13. On a canal boating holiday a family passed through 200 locks. The father kept a record of the rise or fall of the barge at each lock. The frequency distribution below is a copy of his records.

Rise or fall in metres	2.0	2.5	3.0	3.5	4.0	4.5	5.0	5.5	6.0	6.5	7.0	7.5	8.0	8.5	9.0
Frequency	5	10	15	37	56	32	15	12	8	3	2	1	2	1	1

Find the modal lock height.

Mode of Grouped Data

When data has been grouped into class intervals it is difficult to find which value in the data is the mode, but it is a simple matter to find the *modal class* of the data.

The *modal class* is the class interval which turns up the *most*.

Example: The following table shows the results of a Spanish test given to 50 pupils. Group the results in intervals of 1–10, 11–20, 21–30 and then find the modal class.

```
62  48  42  74  50  49  44  70  68  58
18  22   9  18  14  46  45  76  65  70
40  74  50  65  63  82  36  37  39  79
22  14  26  29  32  47  42  98  54  88
19  56  20  16  33  27  16  35  42  59
```

The frequency distribution for the above is:

Mark Ranges	Tallies	Frequency
1–10	I	1
11–20	IIIII III	8
21–30	IIIII	5
31–40	IIIII II	7
41–50	IIIII IIIII I	11
51–60	IIII	4
61–70	IIIII II	7
71–80	IIII	4
81–90	II	2
91–100	I	1

Mode ——→ 41–50 ... 11 ←—— Most

In this case the interval which turns up most is 41–50.
The modal class is 41–50.
The actual mode of grouped data lies within the modal class interval.
It is possible to obtain an *estimate* of the actual mode of grouped data from its

histogram, but only when the intervals on either side of the modal interval have the *same class size* as the modal interval.

To find an *estimate* of the actual mode of grouped data:

(*a*) draw a histogram of the data.
(*b*) find the bar which shows the modal class interval.
(*c*) label the bar from (*b*) with the letters A, B, C and D as shown in the diagram below.

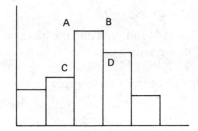
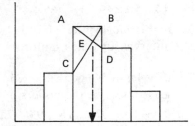

(*d*) join A to D with a straight line.
(*e*) join B to C with a straight line.
(*f*) the point where the lines from (*d*) and (*e*) cut is E.
(*g*) draw a line from E straight down to the horizontal (across) axis.
(*h*) read off the number where the line in (*f*) cuts the horizontal axis.
(*i*) write down the answer to (*h*) as the mode.

Example: Find the modal weight of the frequency distribution below.

Weight kg	50–52	53–55	56–58	59–61	62–64	65–67	68–70
Frequency	2	4	10	12	6	5	1

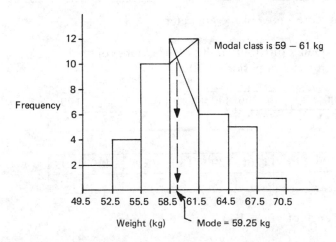

Exercise 4b:

1. The array below shows the times in minutes for which 50 motorists used a set of 5 parking meters in a town centre street.

```
59  52  53  61  27  33  56  20  19  16
58  70  79  88  54  39  65  68  48  22
74  22  40  18  62  26  50   9  42  70
76  37  98  14  74  14  82  42  29  63
46  44  50  49  18  65  45  36  32  47
```

Group this data in intervals of 1–10, 11–20 etc., and then find the modal time interval for which motorists park in the street.

2. The array below shows the times in minutes required by a central heating engineer to service the boilers of 60 of his clients.

```
 90 106 103   76 118   99   72 105 101   73 107   49   95   97   91
145 107   61   98   77 108   88   97   77 110   93   71   65   94 129
 90 121   91   76 114   95   66 115   93   67   92 103   78 111   94
130   89 102 101 114   85   93   77   97 100 113   68 101   79 104
```

Group this data in intervals of 45–59, 60–74, 75–89 etc., and then find the modal time interval that the engineer spends on servicing.

3. The grouped frequency distribution below shows the scores of a cricketer in 40 innings.

Score	0–10	11–20	21–30	31–40	41–50	51–60	61–70	71–80	81–90	91–100	101–110
Frequency	1	2	4	6	9	5	4	3	3	2	1

Find the modal score range of the cricketer.

4. The grouped frequency distribution shows the number of bricks laid per day by a bricklayer over a period of 70 days.

No. of bricks	600–699	700–799	800–899	900–999	1000–1099
Freq. (No. of days)	1	1	2	8	45

No. of bricks	1100–1199	1200–1299	1300–1399	1400–1499
Freq. (No. of days)	9	1	2	1

Find the modal range of bricks laid by the bricklayer.

5. A photographer produced the following frequency distribution which shows the times taken for his flashgun to recharge after each picture he took.

Recharge Time (secs)	0–1	1–2	2–3	3–4	4–5	5–6	6–7	7–8	8–9	9–10
Frequency	2	8	20	36	55	25	10	8	5	1

Find the modal time range needed to recharge the flashgun.

6. The frequency distribution below shows the prices of houses on offer at an estate agents.

Price £'s	16,000–17,000	17,000–18,000	18,000–19,000	19,000–20,000	20,000–21,000	21,000–22,000	22,000–23,000
Freq.	7	18	30	20	10	8	7

Find the modal price range of the houses on offer.

7. During a fishing competition the judges kept a record of the weights of the fish caught. The following distribution shows the weights and frequencies of fish caught during the competition.

Weight g	0–199	200–399	400–599	600–799	800–999	1000–1199	1200–1399	1400–1599
Frequency	10	35	56	40	30	19	6	4

Find the modal weight range of the fish caught.

8. A man who went to work by train noted the number of times the train was late arriving at his destination. The following distribution shows the results of his analysis.

Number of minutes late	0–2	3–5	6–8	9–11	12–14	15–17	18–20
Frequency	5	16	10	7	5	4	3

Find the modal time range of the number of minutes late.

9. Discuss who would be likely to find modal values very useful.

10. Find an *estimate* of the actual mode in each of questions 1 to 8.

11. Use the data you collected about your class to find the:

(a) modal age

(b) modal height
(c) modal waist size
(d) modal chest size
(e) modal hip size
(f) modal weight
(g) modal shoe size,

of the pupils in your class.

Would you expect any of the above to be bi-modal? If so, which and why?

12. Repeat question 11 considering only the girls in your class.

13. Repeat question 11 considering only the boys in your class.

14. Each member of the class with an electric kettle at home should boil exactly 1 pint of water and record the time to the nearest second that the kettle takes to boil. Then group the results and find the modal time range taken for the kettles to boil.

15. Each pupil should fill their bath to a depth of exactly 30 cm and record the time taken to the nearest second for the bath to empty. Group the classes results and find the modal time range for the baths to empty.

16. Try to persuade your local supermarket manager to allow the class to survey the weights of a batch of:
(a) sugar bags
(b) eggs
(c) bags of potatoes
(d) bags of flour, etc.
Find the modal weight ranges of each of the items.

17. Measure your normal pulse rate and then run on the spot for 2 minutes. Record how long it takes for your pulse to return to its normal rate. Find the modal time range taken for pulse rates to return to normal for your class.

18. Choose a page of any book which all of the class will have to read (to themselves) at their normal pace. Each member of the class note how long it takes you to read the page. Then find the modal time range taken to read the page.

19. Choose a book of about 120 pages and allocate to each member of the

class about 4 pages. Each pupil make a frequency distribution of the number of letters per word in your 4 page section. Combine the results of the survey into one distribution for the whole book. Find the modal number of letters per word in the book.

20. Collect as many examples of published statistics as possible and find the modal values contained in the statistics.

Mode from Diagrams
The mode can be found from a diagram even if the diagram is given without a frequency distribution.
The table below shows how the mode can be found from diagrams.

Type of Diagram	Pie Chart	Line Chart	Bar Chart	Pictogram	Histogram
Find the value or interval shown by:	largest sector (slice)	longest line	longest bar	longest line of pictures	largest area

Examples:

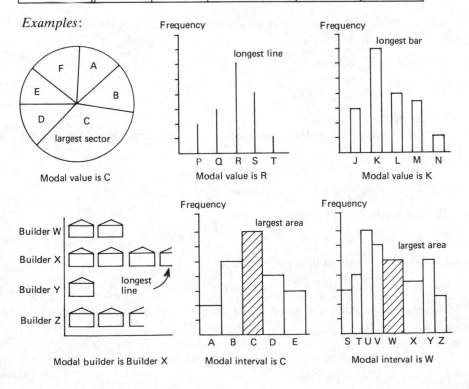

Modal value is C

Modal value is R

Modal value is K

Modal builder is Builder X

Modal interval is C

Modal interval is W

Median

The *median* of a distribution is the *middle* value when all of the data is arranged in numerical order, i.e. order of size.
The median has as many values above as it has below it.
A set of data with an *odd* number of values will have an actual middle value.
A set of data with an *even* number of values will not have an actual middle value. The median is halfway between the two values at the middle of the distribution. The median is the arithmetic mean of the two middle values.

Example: Find the median of 1, 3, 5, 9, 7, 5, 4, 2, 3, 2, 1.
Arranging in numerical order gives:

    1 1 2 2 3 ③ 4 5 5 7 9

                    middle value

The middle value is 3 so the median is 3.

Example: Find the median of 1, 6, 8, 6, 7, 3, 2, 3, 9, 2.
Arranging in numerical order gives:

    1 2 2 3 ③⑥ 6 7 8 9

There is no actual middle value so the median is halfway between the two middle values.

The median is given by $\dfrac{3 + 6}{2} = \dfrac{9}{2} = 4\frac{1}{2}$

Middle of a Set of Data

If an array has X values then the middle value after *arranging in numerical order* will be the $\dfrac{X + 1}{2}$ th value.

If X is *odd* then $\dfrac{X + 1}{2}$ will be a *whole* number, i.e. n.
The median is the n th value.
E.g. If X is 85 then $\dfrac{X + 1}{2} = \dfrac{85 + 1}{2} = \dfrac{86}{2} = 43$
n is 43 so the median is the 43rd value.
If X is *even* then $\dfrac{X + 1}{2}$ will be a *whole* number $+ \frac{1}{2}$, i.e. $p + \frac{1}{2}$.
The median is halfway between the p th and $(p + 1)$th values.
E.g. If $X = 100$ then $\dfrac{X + 1}{2} = \dfrac{100 + 1}{2} = \dfrac{101}{2} = 50\frac{1}{2} = 50 + \frac{1}{2}$.
p is 50 so the median is halfway between the 50th and 51st values.

To find the median of an array with a large number of data:

(*a*) Construct the frequency distribution of the array.
(*b*) Find the Grand Total of the frequencies X and add 1.
(*c*) Divide the answer to (*b*) by 2.
(*d*) Follow the correct side of the table below, depending on the answer to (*c*).

X Odd Whole n	X Even Whole $+\frac{1}{2}$ $p + \frac{1}{2}$
(i) Ring the n th tally down. (ii) Find the value which is along-side the ringed tally. (iii) Write down the answer to (ii) as the median.	(i) Ring the p th tally down. (ii) Ring the $(p + 1)$ th tally down. (iii) Find the values which are alongside the ringed tallies. (iv) Add the answers to (iii) together. (v) Divide the answer to (iv) by 2. (vi) Write down the answer to (v) as the median.

Example: Find the median of the following array.

```
3  1  4  2  1  8  1  6  5  8  9  7  4  6  2
9  3  4  9  5  8  5  4  5  7  4  3  8  2  1
9  1  3  6  4  1  5  7  6  5  7  7  5  8  6
```

The frequency distribution is:

Value	Tallies	Frequency
1	HHT I	6
2	III	3
3	IIII	4
4	HHT I	6
5	HH⊕II	7
6	HHT	5
7	HHT	5
8	HHT	5
9	IIII	4

Ring——→ 5
here

Grand Total X→45

$X = 45$

$X + 1 = 46$

$$\frac{X + 1}{2} = 23$$

$n = 23$

23rd tally down gives 5
median is 5.

Example: Find the median of the following array.

```
3  8  2  6  3  8  6  4  3  7  6  8  1  8  7  5  3  5
3  4  3  7  5  7  9  7  7  7  5  5  8  4  4  7  7  9
8  5  2  9  6  9  6  8  8  4  7  7  8  7  8  6  6  4
5  5  4  6  6  2  3  5  9  9  4  5  7  6  1  8  6  9
```

The frequency distribution is:

Value	Tallies	Frequency
1	II	2
2	III	3
3	HHT II	7
4	HHT III	8
5	HHT HHT	10
both →6	HHT HHT I	11
rings 7	HHT HHT III	13
here 8	HHT HHT I	11
9	HHT II	7

Grand Total $X \rightarrow 72$

$X = 72$

$X + 1 = 73$

$\dfrac{X + 1}{2} = 36\frac{1}{2} = 36 + \frac{1}{2}$

$p = 36$

$p + 1 = 37$

36th tally down gives 6

37th tally down gives 6

$\text{median} = \dfrac{6 + 6}{2} = \dfrac{12}{2} = 6$

Example: Find the median of the following array.

```
5 7 5 5 2 6 6 4 8 5 6 3 6 4 8 5 3 7
6 6 3 6 8 7 1 7 4 9 5 8 4 2 5 4 3 7
4 7 6 9 6 6 2 4 7 4 1 6 8 5 5 3 7 9
```

The frequency distribution is:

Value	Tallies	Frequency
1	II	2
2	III	3
3	HHT	5
4	HHT III	8
one →5	HHT IIII	9
ring →6	HHT HHT I	11
in 7	HHT III	8
each 8	HHT	5
9	III	3

Grand Total $X \rightarrow 54$

$X = 54$

$X + 1 = 55$

$\dfrac{X + 1}{2} = 27\frac{1}{2} = 27 + \frac{1}{2}$

$p = 27$

$p + 1 = 28$

27th tally down gives 5

28th tally down gives 6

$\text{median} = \dfrac{5 + 6}{2} = \dfrac{11}{2} = 5\frac{1}{2}$

Median from a Frequency Distribution
Either:

> (a) rewrite the Frequency Distribution, complete with tallies.
> (b) proceed as described on pages 87–8.

Or:

> (a) make and fill in a cumulative frequency column.
> (b) find the Grand Total of the frequencies X and *add* 1.

(c) divide the answer to (b) by 2.

(d) follow the correct side of the table below depending on the answer to (c).

X Odd Whole *n*	X Even Whole + ½ *p* + ½
(i) Find the number in the cum. freq. column which is either equal to *n*, or is the first cum. freq. in the table which is greater than *n*. (ii) Find the value which has the cum. freq. from (i) alongside it. (iii) Write down the answer to (ii) as the median.	(i) Find the number in the cum. freq. column which is either equal to *p*, or is the first cum. freq. in the table greater than *p*. (ii) Find the number in the cum. freq. column which is either equal to (*p* + 1), or is the first cum. freq. in the table greater than (*p* + 1). (iii) Find the values which have the cum. freq. from (i) and (ii) alongside them. (iv) Add the answers to (iii) together. (v) Divide the answer to (iv) by 2. (vi) Write down the answer to (v) as the median.

Example: Find the median of the frequency distribution below.

Mark	1	2	3	4	5	6	7	8	9
Frequency	1	5	9	14	10	9	4	3	2

Making a cumulative frequency column gives:

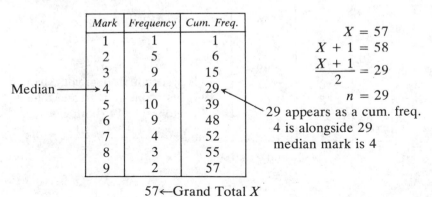

Mark	Frequency	Cum. Freq.
1	1	1
2	5	6
3	9	15
4	14	29
5	10	39
6	9	48
7	4	52
8	3	55
9	2	57

Median ⟶ 4

57 ← Grand Total X

$$X = 57$$
$$X + 1 = 58$$
$$\frac{X + 1}{2} = 29$$
$$n = 29$$

29 appears as a cum. freq.
4 is alongside 29
median mark is 4

Example: Find the median of the frequency distribution below.

Mark	1	2	3	4	5	6	7	8	9
Frequency	0	2	4	6	8	7	3	2	1

Making a cumulative frequency column gives:

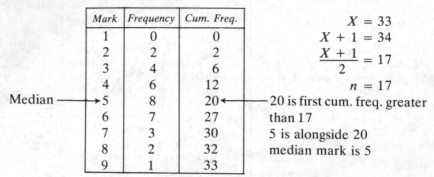

Mark	Frequency	Cum. Freq.
1	0	0
2	2	2
3	4	6
4	6	12
5	8	20
6	7	27
7	3	30
8	2	32
9	1	33

Median ——→ 5 is alongside 20

$X = 33$
$X + 1 = 34$
$\dfrac{X + 1}{2} = 17$
$n = 17$

20 is first cum. freq. greater than 17
5 is alongside 20
median mark is 5

33←Grand Total X

Example: Find the median of the frequency distribution below.

Mark	1	2	3	4	5	6	7	8	9
Frequency	1	3	8	17	20	12	8	2	1

Making a cumulative frequency column gives:

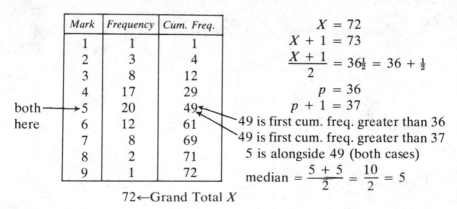

Mark	Frequency	Cum. Freq.
1	1	1
2	3	4
3	8	12
4	17	29
5	20	49
6	12	61
7	8	69
8	2	71
9	1	72

both —→ 5
here

$X = 72$
$X + 1 = 73$
$\dfrac{X + 1}{2} = 36\tfrac{1}{2} = 36 + \tfrac{1}{2}$
$p = 36$
$p + 1 = 37$

49 is first cum. freq. greater than 36
49 is first cum. freq. greater than 37
5 is alongside 49 (both cases)
$\text{median} = \dfrac{5 + 5}{2} = \dfrac{10}{2} = 5$

72←Grand Total X

Example: Find the median of the frequency distribution below.

Mark	1	2	3	4	5	6	7	8	9	10	11	12
Frequency	4	10	16	40	50	30	20	30	15	10	8	7

Making a cumulative frequency column gives:

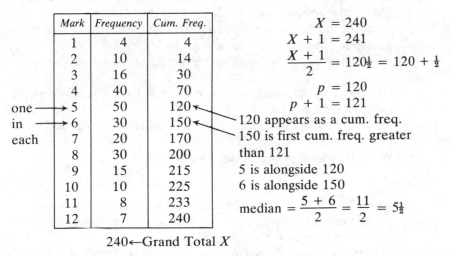

Mark	Frequency	Cum. Freq.
1	4	4
2	10	14
3	16	30
4	40	70
one ⟶ 5	50	120
in ⟶ 6	30	150
each 7	20	170
8	30	200
9	15	215
10	10	225
11	8	233
12	7	240

240←Grand Total X

$X = 240$

$X + 1 = 241$

$\dfrac{X + 1}{2} = 120\tfrac{1}{2} = 120 + \tfrac{1}{2}$

$p = 120$

$p + 1 = 121$

120 appears as a cum. freq.

150 is first cum. freq. greater than 121

5 is alongside 120

6 is alongside 150

$\text{median} = \dfrac{5 + 6}{2} = \dfrac{11}{2} = 5\tfrac{1}{2}$

Exercise 4c:
Repeat all questions in Exercise 4a but in each question find the median value instead of the modal value.

Median of Grouped Data
When data has been grouped, although it is possible to find which class interval the median lies in, it is not possible to find the *actual* median from the cumulative frequency distribution of the data.
The *actual* median of grouped data must be *estimated*.
It is not a straightforward task to estimate the actual median of grouped data.
Since half of the values lie above the median and half lie below it, it is possible to estimate the median from a cumulative frequency curve (ogive) of the data.
The easiest method of estimating the median of grouped data is to use a cumulative frequency curve.
Note: Class boundaries must be used when plotting the cumulative frequency curve(s).

Median from a Cumulative Frequency Curve
Method 1

(a) Draw a cumulative frequency curve of the data.
(b) Find the cumulative frequency on the vertical (up) axis which is the same as the highest cumulative frequency plotted on the curve.
(c) Divide the answer to (b) by 2.
(d) Find the point on the vertical (up) axis which gives the same cumulative frequency as the answer to (c).
(e) Draw a line from this point on the vertical (up) axis *across* to the curve.
(f) Draw a line from where the line in (e) cuts the curve, *down* to the horizontal (across) axis.
(g) Read off the number where the line in (f) cuts the horizontal (across) axis.
(h) Write down the answer to (g) as the median.

Method 2

(a) Draw a cumulative frequency curve of the data to show *less than*.
(b) Draw a cumulative frequency curve of the data to show *more than*, on the same axes as the curve from (a).
(c) Find the point where the curves from (a) and (b) cross each other.
(d) Draw a line from this point *down* to the horizontal (*across*) axis.
(e) Read off the number where the line in (d) cuts the horizontal (across) axis.
(f) Write down the answer to (e) as the median.

Example: Find the median of the frequency distribution below.

Class	1–10	11–20	21–30	31–40	41–50	51–60	61–70	71–80	81–90	91–100
Frequency	2	8	20	50	80	60	40	20	12	8

The cumulative frequency distributions using class boundaries are:

Class	Freq.	C.F.	Less than
1–10	2	2	10.5
11–20	8	10	20.5
21–30	20	30	30.5
31–40	50	80	40.5
41–50	80	160	50.5 ◄──── Class boundaries
51–60	60	220	60.5
61–70	40	260	70.5
71–80	20	280	80.5
81–90	12	292	90.5
91–100	8	300	100.5

Class	Freq.	C.F.	More than
91–100	8	8	90.5.
81–90	12	20	80.5
71–80	20	40	70.5
61–70	40	80	60.5
51–60	60	140	50.5 ◄──── Class boundaries
41–50	80	220	40.5
31–40	50	270	30.5
21–30	20	290	20.5
11–20	8	298	10.5
1–10	2	300	0.5

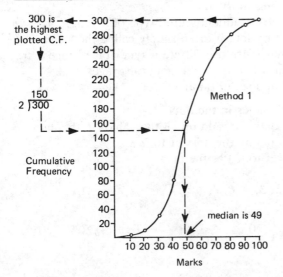

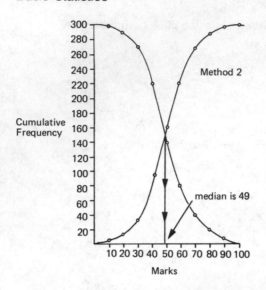

Exercise 4d:
Repeat all questions in Exercise 4b but in each question find the median value instead of the modal ranges.

Arithmetic Mean

The *arithmetic mean* of a set of data is the total of the values in the data divided by the number of values in the data.
The arithmetic mean is the most frequently used type of average.
The arithmetic mean of a set of data is often simply called the *mean*.
People who do not know that there are different types of averages, are usually referring to the mean, when they talk about averages.
To find the *arithmetic mean* of a set of data:

   (*a*) count the number of values in the data.
   (*b*) find the total of the data, i.e. add up all the values in the data.
   (*c*) divide the answer to (*b*) by the answer to (*a*).
   (*d*) write down the answer to (*c*) as the mean.

Example: Find the mean of 5, 3, 4, 6, 2.

(*a*) There are 5 values.
(*b*) 5 + 3 + 4 + 6 + 2 = 20.

(*d*) Mean = 4.

(*c*)
$$\begin{array}{r} 4 \\ 5\overline{)20} \\ \underline{20} \\ \text{--} \end{array}$$

Example: Find the mean of 1, 3, 7, 6, 5, 5, 7, 5, 6, 5.

(*a*) There are 10 values.
(*b*) 1 + 3 + 7 + 6 + 5 + 5 + 7 + 5 + 6 + 5 = 50
(*c*)   5
     10)50
        50
        --
(*d*) Mean = 5.

Example: Find the mean of 21, 12, 11, 17, 21, 25, 18, 15, 27, 37.

(*a*) There are 10 values.
(*b*) 21 + 12 + 11 + 17 + 21 + 25 + 18 + 15 + 27 + 37 = 204.
(*c*)   20.4
     10)204.0
        20
        4
        0
        40
        40
        --
(*d*) Mean = 20.4.

When there is a large amount of data it becomes simpler to find the mean using a frequency distribution of the data.
To find the mean using a frequency distribution:

(*a*) construct a frequency distribution.
(*b*) make a new column to the right of the frequency (totals) column.
(*c*) for each line of the distribution, multiply the value in the 1st column by the frequency of that value.
(*d*) put the answers to (*c*) in the new column, i.e. value × freq.
(*e*) add up the frequencies. This will give the number of values in the data.
(*f*) add up the numbers in the new column. This will give the *total* of the data.
(*g*) divide the answer to (*f*) by the answer to (*e*).
(*h*) write down the answer to (*g*) as the mean.

Example: Find the mean of the following array:

| 1 | 7 | 5 | 1 | 8 | 6 | 5 | 3 | 1 | 4 | 6 | 2 | 8 | 3 | 1 | 5 | 6 | 3 | 4 | 2 |
| 9 | 7 | 4 | 7 | 1 | 9 | 9 | 6 | 10 | 8 | 4 | 6 | 1 | 0 | 8 | 8 | 9 | 3 | 4 | 6 |

The frequency distribution is:

Value	Tallies	Freq.	Value × Freq.
0	I	1	0 × 1 = 0
1	HHT I	6	1 × 6 = 6
2	II	2	2 × 2 = 4
3	IIII	4	3 × 4 = 12
4	HHT	5	4 × 5 = 20
5	III	3	5 × 3 = 15
6	HHT I	6	6 × 6 = 36
7	III	3	7 × 3 = 21
8	HHT	5	8 × 5 = 40
9	IIII	4	9 × 4 = 36
10	I	1	10 × 1 = 10

New column

Number of values = 40
Total of data = 200

$$40 \overline{)200} \quad \begin{array}{r} 5 \\ \underline{200} \\ \text{---} \end{array}$$

Mean = 5.

40   200
Number   Total of Data.
of Values

It is usual to head the first column, i.e. the value column of a frequency distribution, with the letter x.

It is usual to head the frequency column with the letter f.

The value × frequency column then becomes the xf or the fx column.

The *product* of two numbers is the result when the two numbers are multiplied together.

The symbol Σ means 'the sum of' whatever letter(s) follow the symbol.

Σf means 'the sum of all the different f's'.

        or the result of adding up all of the different f's.

        or in this case the total of the frequencies, i.e. the number of values.

Σf can also be written simply as N.

Σfx means 'the sum of all the different products formed from the f's and the x's'.

        or the result of adding up all of the products formed from the f's and the x's.

        or in this case the total of the data, i.e. the total of the value × frequency column.

$$\text{Mean} = \frac{\text{Total of Data}}{\text{Number of Values}} = \frac{\Sigma fx}{\Sigma f} \text{ or } \frac{\Sigma fx}{N}$$

Using the above symbols the previous example becomes:

Value x	Tallies	Freq. f	Value × Freq. fx
0	I	1	$0 \times 1 = 0$
1	IIII I	6	$1 \times 6 = 6$
2	II	2	$2 \times 2 = 4$
3	IIII	4	$3 \times 4 = 12$
4	IIII	5	$4 \times 5 = 20$
5	III	3	$5 \times 3 = 15$
6	IIII I	6	$6 \times 6 = 36$
7	III	3	$7 \times 3 = 21$
8	IIII	5	$8 \times 5 = 40$
9	IIII	4	$9 \times 4 = 36$
10	I	1	$10 \times 1 = 10$

$\Sigma f = 40$ $\Sigma fx = 200$
Number of   Total of Data.
Values

Σf = Number of values = 40
Σfx = Total of Data = 200

$$\begin{array}{r} 5 \\ 40\overline{)200} \\ 200 \\ \hline --- \end{array}$$

$\dfrac{\Sigma fx}{\Sigma f} = 5$

Mean = 5.

Example: Find the mean of the following array.

```
90  88  91  91  92  90  92  87  93  89
92  91  89  88  89  90  88  93  87  92
86  94  89  91  88  90  90  87  93  90
```

The frequency distribution is:

x Value	Tallies	f Freq.	fx Freq. × Value
86	I	1	86
87	III	3	261
88	IIII	4	352
89	IIII	4	356
90	IIII I	6	540
91	IIII	4	364
92	IIII	4	368
93	III	3	279
94	I	1	94

$\Sigma f = 30$ $\Sigma fx = 2700$

Number of values = Σf = 30
Total of data = Σfx = 2700

$$\begin{array}{r} 90 \\ 30\overline{)2700} \\ 270 \\ \hline ---0 \end{array}$$

$\dfrac{\Sigma fx}{\Sigma f} = 90$

Mean = 90.

Example: Find the mean of the frequency distribution below.

Number	50	52	54	56	58	60	62	64	66	68	70
Freq.	2	5	4	5	15	20	15	16	10	5	3

Rewriting the frequency distribution gives:

x	f	fx
50	2	100
52	5	260
54	4	216
56	5	280
58	15	870
60	20	1200
62	15	930
64	16	1024
66	10	660
68	5	340
70	3	210

$\Sigma f = 100$

$\Sigma fx = 6090$

$\dfrac{\Sigma fx}{\Sigma f} = 60.9$

Mean $= 60.9$

$$\begin{array}{r} 60.9 \\ 100\overline{)6090.0} \\ 600 \\ \hline --90 \\ 0 \\ \hline 900 \\ 900 \\ \hline --- \end{array}$$

$\Sigma f = 100$

$\Sigma fx = 6090$

Exercise 4e:
Repeat all questions in Exercise 4a but in each question find the mean value instead of the modal value.

Mean of Grouped Data
When data has been grouped it is not possible to find the *actual* mean of the data.
The *actual* mean of grouped data must be *estimated*.
It seems reasonable to speculate that the values in any class interval will be equally spread out around the mid-value of the class interval.
If this speculation is true, then the mid-value of each class interval can be taken to represent the class as a whole.
This speculation is frequently not true, but using the mid-value of each interval to represent the interval gives sufficiently *accurate* results to justify the use of mid-values.

To find the mid-value of a class interval:

(a) add the lower class limit of the interval to the upper class limit of the interval.

(b) divide the answer to (a) by 2.

Examples:

Interval	a		b	Mid-Value
1–10	1 + 10	= 11	11 ÷ 2 = 5.5	5.5
0–9	0 + 9	= 9	9 ÷ 2 = 4.5	4.5
1–5	1 + 5	= 6	6 ÷ 2 = 3	3
50–54	50 + 54	= 104	104 ÷ 2 = 52	52
20.5–22.5	20.5 + 22.5	= 43	43 ÷ 2 = 21.5	21.5
600–900	600 + 900	= 1500	1500 ÷ 2 = 750	750
43.5–48.5	43.5 + 48.5	= 92	92 ÷ 2 = 46	46

To find an estimate of the actual mean of grouped data:

(a) construct a frequency distribution with two new columns.

(b) find the mid-value of each interval and place them in the 1st new column i.e. mid-value.

(c) for each line of distribution, multiply the mid-value by the frequency of that mid-value.

(d) put the answers to (c) in the second new column, i.e. freq. × mid-value.

(e) add up the frequencies. This gives the number of values in the data.

(f) add up the second new column. This will give an estimate of the *total* of the data.

(g) divide the answer to (f) by the answer to (e).

(h) write down the answer to (g) as the mean.

The mid-value column can be headed with the symbol X.

Example: Find the mean of the following array by grouping in class intervals of 1–5, 6–10, 11–14, etc.

```
34  18  38  29  43  21  23  29  19  34
36  33  27  27   3  31  13  27  40  26
17  37  27  32  26  25  39  23  26  32
23  30  20  35   8  12  28  35  26  32
22  30  28  42  47  24  44  16  30  48
```

The frequency distribution is:

Class	Tallies	Freq.	Mid-Val.	Freq × Mid-Val.
1– 5	I	1	3	3
6–10	I	1	8	8
11–15	II	2	13	26
16–20	̶H̶H̶	5	18	90
21–25	̶H̶H̶ II	7	23	161
26–30	̶H̶H̶ ̶H̶H̶ ̶H̶H̶	15	28	420
31–35	̶H̶H̶ IIII	9	33	297
36–40	̶H̶H̶	5	38	190
41–45	III	3	43	129
46–50	II	2	48	96

$\Sigma f = 50$
$\Sigma fX = 1420$

$$\begin{array}{r} 28.4 \\ 50\overline{)1420.0} \\ 100 \\ \hline 420 \\ 400 \\ \hline 200 \\ 200 \\ \hline \text{- - -} \end{array}$$

$\Sigma f = 50$ $\Sigma fX = 1420$ $\dfrac{\Sigma fX}{\Sigma f} = 28.4$

Number of Values   Total of Data   Mean = 28.4

Note: The estimate of the mean using mid-values is 28.4. The actual mean found by adding up all of the data in the array is 28.3.

Example: Find the mean of the frequency distribution below.

Class	0–24	25–49	50–74	75–99
Freq.	17	49	46	8

Rewriting the frequency distribution gives:

Class	f	X	fX
0–24	17	12	204
25–49	49	37	1813
50–74	46	62	2852
75–99	8	87	696

$\Sigma f = 120$ $\Sigma fX = 5565$

$\Sigma f = 120$
$\Sigma fX = 5565$
$\dfrac{\Sigma fX}{\Sigma f} = 46.375$
Mean $= 46.375$

$$\begin{array}{r} 46.375 \\ 120\overline{)5565.000} \\ 480 \\ \hline 765 \\ 720 \\ \hline 450 \\ 360 \\ \hline 900 \\ 840 \\ \hline 600 \\ 600 \\ \hline \text{- - -} \end{array}$$

Exercise 4f:
Repeat all questions in Exercise 4b but in each question find the mean *value* instead of the modal ranges.

Averages

Comparison of Mode, Median and Mean.

Mode

Advantages	Disadvantages
1. It is easy to understand. 2. Very high and very low values do not affect it. 3. It can be used for finding the averages of items which are not easily measured mathematically e.g. colours of socks. 4. It is a value contained in the original data. 5. It can be found if all of the data is not available: only the most frequent values need to be known.	1. Finding the mode of grouped data is difficult. 2. It is not suitable for use in the calculation of other statistical quantities. 3. Bi-modal, tri-modal distributions occur.

Median

Advantages	Disadvantages
1. It is easy to understand. 2. Very high and very low values do not affect it. 3. It is normally found easily. 4. It is usually a value which is contained in the original data. 5. It can be found if all the data is not available: only the middle values need be known.	1. If there are only a few values it is unlikely to give a good representative average. 2. Arranging the data in order and finding the middle value can be slow and tedious. 3. Finding the median of grouped data is difficult. 4. It is not suitable for use in the calculation of other statistical quantities.

Mean

Advantages	Disadvantages
1. It is commonly used. 2. With practice it is easy to calculate. 3. With study it is easy to understand. 4. It makes use of all the data available. 5. It can be calculated when only the total of the data and the number of values in the data are known. 6. It can be calculated with great accuracy.	1. It is frequently a number which is not in the original table of data. 2. It may be affected a great deal by a few very high or very low values in the data. 3. All of the data or the totals of both the data and the frequencies must be available.

Miscellaneous Exercises 4

Questions 1–5 concern the following numbers:

    1   8   3   4   3   5   8   10   3

1. Find the sum of the numbers.

2. Calculate the arithmetic mean.

3. Write down the median.

4. Write down the mode.

5. Find the range.   EA78

6. At a certain stage in the season 1975–76, the points scored by nine football clubs in the English Second Division were:

    50   50   46   46   43   41   41   41   40

Calculate:
(*a*) the mode.
(*b*) the median.   Y77

7. Find the median of:

    31   57   32   59   42   43   WM75

8. Find the mean, median and mode of the following eleven numbers:

24 19 27 20 19 23 25 19 20 25 21   M78

9. 1 3 4 6 8 3 8 9 3

For the numbers given above determine:
 (i) the median.
 (ii) the mode.
 (iii) the mean.   EM78

10. The following set of numbers give the marks gained by a group of children in a test:

2 8 4 6 2 3 3 5 9 5 6 6 9 3 4 4 5 5 7 6 7 4 5 6 6

(*a*) Find the mode and the median of this set.
(*b*) Calculate the arithmetic mean of this set.
(*c*) What percentage of the children scored more than 5 marks? WM76

11. Find the arithmetic mean of:

120   121   121   124   129   WM75

12. Ten electric bulbs were tested until they burnt out. The following numbers give the life in hours of the bulbs.

19   25   30   21   21   24   25   25   22   24

(*a*) For this set of results, find:
   (i) the arithmetic mean;
   (ii) the mode.
(*b*) When one extra bulb is tested the arithmetic mean for the whole group of 11 bulbs is 24 hours. What was the life of the extra bulb?
     WM77

13. Find the mean, median and mode of the following seven numbers:

9.1   9.3   9.7   9.9   9.1   9.5   9.2   M77

14. The marks scored in a skating contest were:

26.65   28.50   21.25   43.75   33.40

(*a*) Rearrange these figures in ascending order of size.
(*b*) For the above marks find:
   (i) the median,
   (ii) the arithmetic mean.   Y78

15. Find the mean of the six numbers written below:

37   43   49   101   31   51   Y78

16. What is the mean of the numbers:

7   9   11   8   10?   Y77

17. Find the arithmetic mean of:

9900   9901   9903   9905   9906   9909   WM76

18. Find the arithmetic mean of:

181.4   179.4   178.9   177.4   180.9   180.2   WM77

19. Find the arithmetic mean of:

251.2   249.8   248.7   251.7   249.1   WM78

20. The following are the heights in cm to the nearest cm of the 20 children in class 5R:

180   176   176   184   176   176   186   187   161   169
178   184   186   185   178   176   161   190   175   176

(*a*) Calculate the mean of these heights.
(*b*) Find and write down the median height.
(*c*) Write down the modal value of the heights.
(*d*) What percentage of the children are taller than 182 cm?   Y77

21. In 1976, a leading British car company manufactured 688,800 cars with a workforce of 120,000. Calculate the mean (average) number of cars produced by each member of the workforce.   Y78

22. The following table gives the distribution of scores by candidates in an examination.

Score	0–9	10–19	20–29	30–39	40–49	50–59
Frequency	8	22	63	82	86	91

Score	60–69	70–79	80–89	90–99
Frequency	66	53	22	7

(*a*) Calculate the total number of candidates.
(*b*) Calculate the number of candidates who scored:
   (i) not less than 70 marks;
   (ii) more than 19 marks but less than 60 marks.
(*c*) Find:
   (i) the modal class;
   (ii) the class containing the median score.

(*d*) The class intervals are varied. Copy and complete each of the following tables.

(i)

Score	0–19	20–49	50–69	70–99
Frequency				

(ii)

Score	50–54	55–59	60–64	65–69
Frequency		42	38	

WM77

23. The number of goals scored by each of the 22 teams in Division 1 of the Football League in season 1973–4 is listed below.

66 52 52 67 54 49 56 55 49 45 49
44 39 50 51 52 47 38 37 56 43 56

(i) Tabulate these figures to form a grouped frequency distribution, starting at 35 and taking class intervals of 5.
(ii) What is the modal class interval?
(iii) Using the frequencies obtained and the mid-values of the class intervals, present this information in the form of a frequency polygon.   L78

24. (*a*) The heights, in cm, of 32 children are given below.

151 136 150 154 142 144 152 141
149 142 157 148 146 156 138 150
145 163 159 146 147 152 158 152
153 147 148 151 158 149 145 146

The table below is a frequency table, which, when completed, will indicate the number of children whose height is in the ranges shown. (You can assume that all the range intervals are equal.)
Copy the table and complete *all* the blank spaces in it and once you have completed the table state which is the modal class of the ranges of heights given, explaining how you obtained your answer.

Range of heights (cm)	Frequency of heights
135–139	2
	4
145–149	
155–159	

(b) The numbers of pence pocket money received by 25 children in a class, in a recent week, are given below.

60	120	40	35	100
100	45	50	70	125
55	150	50	75	80
65	80	100	30	50
75	50	40	110	75

(i) In this recent week, how many children had:
   (1) 75p or more pocket money
   (2) 50p or less pocket money?
(ii) Find the median and the range of the above 25 numbers of pence, explaining briefly how you obtained each answer. M77

25. (a) The marks gained by 25 pupils in a spelling test are given below.

15	13	11	20	19
18	6	17	12	12
16	17	20	14	13
14	13	12	11	9
12	15	12	13	10

For this set of marks state:
   (i) the range   (ii) the median   (iii) the mode

explaining how you obtained each answer.

(b) The weekly wages (in £) of 30 people are given below.

50	82	75	62	83	61	76	44	71	64
83	64	82	72	56	66	65	84	66	75
72	83	65	61	76	68	88	73	53	65

The table below is a frequency table, which, when completed, will indicate the number of people whose wage is in the ranges shown. (You can assume that all the range intervals are equal.)
Copy the table and complete *all* the blank spaces in it. Use your completed table to state which is the modal class of the ranges of wages given, explaining how you obtained your answer.

Range of Wages(£)	Number of people
40–49	1
	3
60–69	
80–89	

26. The numbers below are the marks of 30 pupils, 16 boys and 14 girls, in a test.

Boys				Girls			
5	3	8	4	1	5	7	5
6	4	5	0	8	5	6	6
10	2	5	4	6	3	9	5
5	6	7	6	6	7		

A table has been started below showing the number of pupils who got marks of 10, 9 and 8. Copy out this table and continue it for marks 7, 6, 5, 4, 3, 2, 1 and 0. Fill in *all* the blank spaces in your table.

Mark	Total Number Getting the Mark		
	Boys	Girls	All pupils
10	1	0	1
9	0	1	1
8	1	1	2

Find the answers to the questions which follow:

(*a*) How many boys got 5 marks or less?
(*b*) How many girls got 6 marks or more?
(*c*) How many pupils got less than 4 marks?
(*d*) When considering the marks for all the pupils, what is:
    (i) the mean,
    (ii) the median,
    (iii) the mode?

(Explain *carefully* how you obtained the answers to (ii) and (iii).)
M76

27. In a swimming match, the times taken, to the nearest second, by 20 children to swim a length were as follows:

31 27 24 26 31 25 26 32 27 31
26 32 30 32 29 25 29 27 26 32

(*a*) Complete the table given below.

Time (s)	Frequency	Time × Frequency
24		
25		
26		
27		
28		
29		
30		
31		
32		
Totals		

(*b*) Find the mean time taken, giving your answer to three significant figures.

(*c*) Find the median time taken. Y77

28. A class recorded their shoe sizes and obtained the following distribution:

Shoe size	3	4	5	6	7	8
No. of children	5	9	8	5	1	4

(*a*) How many children were there in the class?
(*b*) Write down the mode of this distribution.
(*c*) Calculate the arithmetic mean of this distribution. WM78

29. In a traffic census the number of passengers (excluding the driver) being carried in a motor car was counted. The table shows the percentage of motor cars with 0, 1, 2, 3, 4 or 5 passengers.

No. of Passengers	0	1	2	3	4	5
Percentage of Motor Cars	57.6	30.1	5.1	4.0	2.1	1.1

Calculate the mean number of passengers per car. EM78

30.

No. of Houses	Weekly Rental
45	£5
35	£4
20	£2.50

The above figures show the weekly rent on houses on an estate.
Calculate the mean weekly rent per house. EA78

31.

```
5  1  7  7  2  2  4  4  3  3
4  5  9  3  9  3  5  8  6  3
5  8  5  6  4  9  3  7  1  5
6  3  9  4  4  8  2  1  6  3
1  9  1  9  7  5  5  3  2  7
```

Find the mean value of the 50 numbers given in the array above. WM78

32. Draw up a grouped frequency table for the numbers given in question 31, using the classes shown below.

Classes	Tally marks	Frequency
1–3		
4–6		
7–9		

WM78

33. Calculate the mean value of the numbers from the grouped data assembled in question 32. WM78

34. Why would you expect the answers to questions 31 and 33 to be different? WM78

35. (*a*) Find the sum and the mean of the following six numbers

   17 20 13 19 18 15

(*b*) Find the mean of the six numbers in part (*a*) when each of the numbers is increased by one hundred. M77

36. The arithmetic mean of 6 numbers is 16 and of a further 5 numbers is 5. Calculate the mean of all 11 numbers. Y78

37. The average weekly pocket money for 15 boys is 80p and for 10 girls is 90p. What is the average weekly pocket money for these 25 children? M77

38. At a certain seaside town the average number of hours sunshine for the first 20 days in June was 8.6 hours per day. If the average for the whole month was 8.8 hours per day, find the average for the remaining 10 days.
M78

39. (*a*) In a Mathematics test five pupils scored 81, 31, 47, 49 and 62 respectively. Calculate the mean score.

(*b*) Four girls weigh 41.3 kg, 38.2 kg, 40.6 kg, and 44.3 kg respectively. When a fifth girl joins them, the mean weight of the five girls is 41.0 kg. Calculate the weight of the fifth girl.
M78

40. (*a*) In a certain district, the average rainfall during the first eight months of the year was 6.0 cm per month, whilst the average for the last four months was 4.8 cm per month. What was the average rainfall per month for the whole year?
M78

41. The mean height of four pupils is 165 cm. The mean height of three of these pupils is 162 cm. Calculate the height of the fourth pupil.   Y77

42. A sample of 9 numbers has a mean of 4.3. Another sample of 5 numbers has a mean of 5.7. If the two samples are put together what is the mean of the sample of 14 numbers?
WM76

43. A proficiency test used by a company to assess applicants for employment yields the following scores for 100 applicants.

Score	0–4	5–9	10–14	15–19	20–24	25–29	30–34	35–39
Frequency	3	6	22	30	21	10	5	3

(*a*) Calculate the mean of the grouped data by using mid-interval values of 2, 7, 12, etc.

(*b*) Draw a cumulative frequency diagram for the data and use the diagram to find the median. Indicate with the letter *M* where this result is taken on the curve. (Use a scale of 1 cm to 5 marks and 1 cm to 10 applicants.)
M78

44. In a field study of 100 rose bushes during the summer of 1976 the number of perfect blooms grown on each bush was noted. The frequency distribution of the perfect blooms is shown in the following table.

No. of perfect blooms per bush	5–7	8–10	11–13	14–16	17–19	20–22	23–25
Frequency	5	20	30	25	10	7	3
Cumulative frequency	5	25	?	?	?	?	100

(a) Calculate the mean of the grouped data by using mid-interval values of 6, 9, 12, etc.

(b) Copy and complete the table showing the cumulative frequency distribution only.

(c) Draw a cumulative frequency diagram for the data and use your diagram to find the median number of perfect blooms per bush. Indicate with the letter M, the point *on* the curve where this result is found. (Use the upper class limits 7, 10, 13, etc. for this diagram and a scale of 1 cm to 10 perfect blooms on the frequency axis.) M77

45. The table below shows the frequency distribution of the number of passengers per journey on 250 train journeys (all on the same route).

Number of passengers per journey	Number of journeys
1– 5	3
6–10	9
11–15	12
16–20	18
21–25	20
26–30	30
31–35	44
36–40	38
41–45	27
46–50	20
51–55	13
56–60	9
61–65	5
66–70	2

(a) Construct a table showing the cumulative frequency distribution and draw the graph of the cumulative frequency distribution (i.e. the ogive).

(b) Estimate from your graph, indicating clearly how you have obtained each result,

(i) the median number of passengers per journey

(ii) the percentage of journeys on which there were 38 passengers or less.

(c) Check, by calculation and without reference to your graph, the accuracy of your two answers. Comment on any assumptions you make.

M78

46. A small firm decides to manufacture shoes. It has only sufficient capital to install machinery to manufacture one size. In deciding which size to manufacture, is the firm likely to be most interested in the arithmetic mean, the median or the mode of sizes of shoes worn by the population?

EA78

47. State *briefly* the reason for your answer to question no. 46. EA78

48. If you are given 49 numbers, not in numerical order, explain how you would find the median of these numbers. WM78

49. The following is a record of marks obtained by a group of pupils in an examination.

```
24  62  51  57  11  56  64  37  46  34  45
55  36  69  46  51  43  44  47  49   2  16
26  43  39  54  64  61  84  49  28  59  49
52  58  69  57  32  32  27  41  44  53  57
49  64  61  58  39  34  35  47  58  52  76
```

(a) Construct a table showing the frequency distribution in the form '0–10', '11–20', etc.

(b) Use the information to draw a *cumulative frequency curve*.

(c) From the curve estimate the median mark. NW76

50. The following is a record of scores obtained by 8 rugby teams in 10 matches.

```
13  29  35  56  21  26   9  21  20  47
29   7  25  17  19  32  33  18  13  42
24  18  21  19  30  39  12  16  43  30
14  82   3  30  14  40  20  72   8  29
25  14  31  38  46   7  26  16  46  36
25  41  13   9  30  17  32   2  32  18
17   8  27  34  20  68  54  11  12  53
 9  59  18  30  14  20   7  48  39  24
```

(*a*) Construct a table showing the frequency distribution in the form '0–9', '10–19', '20–29', etc.

(*b*) Use the table to draw a bar-chart.

(*c*) In which group does the median lie?   NW77

51. A class of 20 pupils has 12 boys and 8 girls. The average weight of the boys is 69 kg and the average weight of girls is 57 kg. Calculate:

(*a*) the total weight of the boys,

(*b*) the total weight of the girls,

(*c*) the average weight of the class.   NW77

52. The following is a table of merit marks obtained by 80 pupils in a competition.

```
31  64  42  86  19  43  25  60   3  44
53  14  33  50  59  38  65  27  71  30
42  63  60   5  24  77  39  97  15  73
75  44  55  37  20  41  25  43  54  40
21  48  26  39  50  66  51  37  70  29
38  65  88  20  49  49   9  51  38  31
62  55  32  32  29  82  66  39  59  52
50  22  51  74  80  54  33  62  49  48
```

(*a*) Construct a table showing the *cumulative* frequency distribution in the form '0–10', '11–20', '21–30', etc. Use the information to draw a *cumulative frequency curve*.

(*b*) From your curve estimate the median mark.   NW78

53. The following tally chart is the result of a survey at a small factory in which employees were asked how much they earned each week.

Weekly Wage £	Tally
15	HHI I
25	III
35	HHI HHI IIII
45	HHI HHI HHI HHI HHI III
55	HHI IIII
65	HHI III
75	III
85	II
95	II

Calculate the mean weekly wage. Give your answer correct to the nearest penny.   NW78

54. Frequency table of House Competition Points.

Points	Frequency
60–64	1
55–59	3
50–54	2
45–49	5
40–44	8
35–39	10
30–34	7
25–29	4
20–24	3
15–19	1
10–14	1
5–9	0

Calculate the mean of the above distribution correct to one decimal place.   NW77

5 DISPERSION

The average of a set of data gives a measure of the central location of the data.

It is unwise to attempt to compare two different sets of data using their averages alone, since the average does not give any indication of how spread out or variable the data is.

Two sets of data can have exactly the same average value, but can be completely different in the way they are spread out or dispersed about that average value.

E.g. The mean of 98, 99, 100, 101, 102 is 100.

The mean of 0, 50, 100, 150, 200 is also 100.

Both arrays have the same mean of 100, but it is obvious that the 2nd array is much more spread out than the 1st array.

It is useful to have a measure of the spread or dispersion or variability of a set of data.

It is possible to compare two sets of data of the same kind if we have both an average and a measure of dispersion for each set of data.

The four most important measures of dispersion are:

1. range
2. semi-interquartile range
3. mean deviation
4. standard deviation.

Each one of the four measures of dispersion has its own:

(a) uses
(b) advantages
(c) disadvantages.

Range

The *range* of a set of data is the difference between the highest value in the data and the lowest value.

To find the range of a set of data:

(a) find the highest value in the data.
(b) find the lowest value in the data.
(c) subtract the answer to (b) from the answer to (a)
(d) write down the answer to (c) as the range.

Examples: Find the range of
98 99 100 101 102

  (*a*) highest value = 102
  (*b*) lowest value = 98
  (*c*) 102 − 98 = 4
  (*d*) range = 4.

Find the range of
0 50 100 150 200

  (*a*) highest value = 200
  (*b*) lowest value = 0
  (*c*) 200 − 0 = 200.
  (*d*) range = 200.

Since the range is dependent only on the highest and lowest values and does not take into account any of the middle values, its usefulness is limited. It is too easily affected by very high or very low values.

Semi-Interquartile Range (S.I.Q.R.)

The median of a set of data divides the data into two equal parts. The quartiles of a set of data divide the data into four equal parts.

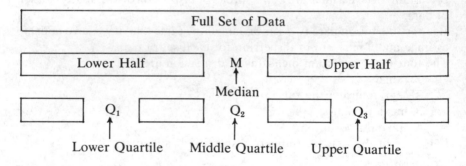

There are three quartiles, these being Q_1, the first or lower quartile, Q_2, the second or middle quartile, and Q_3, the third or upper quartile. The second quartile is in fact the median.

The *semi-interquartile range* is a half of the difference between the upper and lower quartiles.

To find the semi-interquartile range of a set of data:

  (*a*) find the median of the whole set of data.
  (*b*) find the upper quartile by finding the median of the upper half of the data.
  (*c*) find the lower quartile by finding the median of the lower half of the data.
  (*d*) subtract the answer to (*c*) from the answer to (*b*).
  (*e*) divide the answer to (*d*) by 2.
  (*f*) write down the answer to (*e*) as the S.I.Q.R.

Examples: Find the S.I.Q.R. of
98   99   100   101   102

(a) $\boxed{98 \quad 99}$ 1$\underset{\uparrow}{0}$0 $\boxed{101 \quad 102}$

median

(b) 101$\underset{\uparrow}{}$ 102←upper half of data.

median$=\dfrac{101+102}{2}=\dfrac{203}{2}=101\frac{1}{2}$

(c) 98$\underset{\uparrow}{}$ 99←lower half of data.

median$=\dfrac{98+99}{2}=\dfrac{197}{2}=98\frac{1}{2}$

(d) $101\frac{1}{2}-98\frac{1}{2}=3$
(e) $3\div2=1\frac{1}{2}$
(f) S.I.Q.R.$=1\frac{1}{2}$.

Find the S.I.Q.R. of
0   50   100   150   200

(a) $\{0 \quad 50\}$ 1$\underset{\uparrow}{0}$0 $\boxed{150 \quad 200}$

median

(b) 150$\underset{\uparrow}{}$ 200←upper half of data

median$=\dfrac{150+200}{2}=\dfrac{350}{2}=175$

(c) 0$\underset{\uparrow}{}$ 50←lower half of data

median$=\dfrac{0+50}{2}=\dfrac{50}{2}=25$

(d) $175-25=150$
(e) $150\div2=75$
(f) S.I.Q.R.$=75$.

The *lower quartile* is the value which has one quarter of the values below it and three quarters of the values above it.

The *upper quartile* is the value which has one quarter of the values above it and three quarters of the values below it.

The quartiles of a frequency distribution can be found from a cumulative frequency curve or ogive.

To find the upper and lower quartiles of a frequency distribution:

(a) draw a cumulative frequency curve showing 'less than' (see pages 56–7).
(b) find the cumulative frequency on the vertical (up) axis, which is the same as the highest cumulative frequency plotted on the curve.
(c) divide the answer to (b) by 4.
(d) multiply the answer to (c) by 3.
(e) find the points on the vertical (up) axis which give the same cumulative frequencies as the answers to (c) and (d).
(f) draw a line from each of these points on the vertical (up) axis *across* to the curve.
(g) draw a line from where each of the lines in (f) cut the curve, *down* to the horizontal (across) axis.
(h) read off the numbers where the lines in (g) cut the horizontal (across) axis.
(i) write down the left hand number as the lower quartile.
        the right hand number as the upper quartile.

To find the S.I.Q.R. of a frequency distribution use the values obtained from the above method in the formula:

$$S.I.Q.R. = \tfrac{1}{2} \text{ (Upper Quartile − Lower Quartile)}$$

Example: Find the upper and lower quartiles and hence the semi-interquartile range of the frequency distribution below.

Mark	1–10	11–20	21-30	31–40	41–50	51–60	61–70	71–80	81–90	91–100
Frequency	1	5	21	35	35	21	6	3	2	1

Constructing the cumulative frequency distribution we get:

Marks	f	Cum. Freq.	Less than	
1–10	1	1	10.5	
11–20	5	6	20.5	
21–30	21	27	30.5	see the graph below.
31–40	35	62	40.5	
41–50	35	97	50.5	
51–60	21	118	60.5	
61–70	6	124	70.5	
71–80	3	127	80.5	class boundaries
81–90	2	129	90.5	
91–100	1	130	100.5	

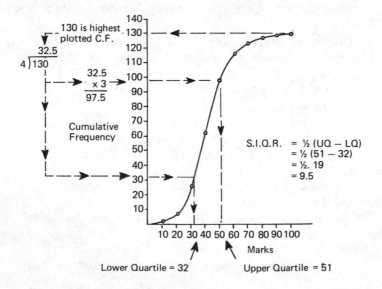

Note: Half (50%) of the values in any set of data lie between the lower and upper quartiles.

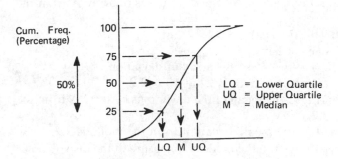

The S.I.Q.R. is dependent only on the middle values of the data and is not affected by very high or very low values.
The usefulness of the S.I.Q.R. is limited.

Mean Deviation

If a number is subtracted from another number the result is called the *difference* between the two numbers.
The difference between V and U is V − U.
E.g. the difference between 5 and 2 is 3 since 5−2=3.
     the difference between 3 and 7 is −4 since 3 −7 = −4.
If the same number, Y, is subtracted from each value in a set of data, then the differences between each of the values in the set and the number Y can be called the differences of the values from the number Y.
Consider the array   5   4   6   8   3.
Subtracting 5 from each value in the array, gives an array of differences which is   0   −1   1   3   −2.
The differences   0   −1   1   3   −2   are the differences of the values in the original array from 5.
If all of the numbers in an array of differences are taken as positive, i.e. minus signs are ignored, then the array of positive differences is called an array of *deviations*.
The array of deviations for the above example is   0   1   1   3   2.
For any array we can find the deviations of that array from the average value of the array.
For any array we can find the deviations from the mode
                    or the deviations from the median
                    or the deviations from the mean.

Consider array A which is 1 2 2 2 3 3 5 6 6 10.
The three averages of array A are mode = 2, median = 3, mean = 4.
If we subtract the mode i.e. 2, from every value in array A, we obtain an array B, which is an array made up of the differences of array A from the mode of array A.
Array B is −1 0 0 0 1 1 3 4 4 8.
Ignoring the minus sign we have
     1 0 0 0 1 1 3 4 4 8←array of deviations.
The mean of this array of deviations is 22 ÷ 10 = 2.2.
We can say that for array A the mean of the deviations of array A from the mode of array A is 2.2.
This can be shortened to the mean deviation from the mode is 2.2.
The mean of the deviations from thc mode is a measure of the dispersion or spread of the values in the data about the mode.
We can find the mean of the deviations from the median or from the mean.
Consider array A again. 1 2 2 2 3 3 5 6 6 10
Subtracting the median i.e. 3, from array A gives:
    −2 −1 −1 −1 0 0 2 3 3 7
This is an array of differences from the median.
Ignoring the minus signs gives:
    2 1 1 1 0 0 2 3 3 7←deviations
The mean of this array of deviations is 20 ÷ 10 = 2.
The mean deviation from the median is 2.
The mean deviation from the median is a measure of dispersion about the median.
Subtracting the mean, i.e. 4, from array A gives:
    −3 −2 −2 −2 −1 −1 1 2 2 6
This is an array of differences from the mean.
Ignoring the minus signs gives:
    3 2 2 2 1 1 1 2 2 6←deviations
The mean of this array of deviations is 22 ÷ 10 =2.2.
The mean deviation from the mean is 2.2.
The mean deviation from the mean is a measure of dispersion about the mean.

Note: Although the mean of the deviations can be found from the mode or the median or the mean, it is usual to find only the mean deviation from the mean.

Examples:

Find the mean deviation of
98 99 100 101 102
Mean = 100

Find the mean deviation of
0 50 100 150 200
Mean = 100

Subtracting the mean gives
−2 −1 0 1 2
Ignoring the minus signs
2 1 0 1 2
mean of deviations = 6 ÷ 5 =1.2
Mean deviation from the mean=1.2.

Subtracting the mean gives
−100 −50 0 50 100
Ignoring the minus signs
100 50 0 50 100
mean of deviations = 300 ÷5 = 60
Mean deviation from the mean=60.

$$\text{Mean Deviation} = \frac{\text{Total of the Deviations}}{\text{Number of Values}}$$

The mean of a set of data can be denoted by $\bar{x}$, in the case of ungrouped data, and $\overline{X}$ in the case of grouped data.

Note: The terms *differences* and *deviations* are often taken to have the same meaning.

To find the mean deviation from the mean of a frequency distribution:

(a) construct the frequency distribution and using the methods described on pages 95–100 find the mean of the distribution.

(b) make three new columns to the right of the table obtained when step (a) has been completed.

For ungrouped data:

x Value	Tallies	f Freq.	fx Freq. × Val.		$\bar{x}$ Mean	$x-\bar{x}$ Deviation Val. − Mean	$f(x-\bar{x})$ Freq. × Deviation Freq. × (Val. − Mean)

For grouped data:

Interval	Tallies	f Freq.	X Mid. Val.	fX Freq. × M.V.		$\overline{X}$ Mean	$X-\overline{X}$ Deviation M.V. − Mean	$f(X-\overline{X})$ Freq. × Deviation Freq. × (M.V. − Mean)

(c) put the mean $\bar{x}$ or $\overline{X}$ worked out in (a) on every line of the first new column i.e. mean.

(d) for each line of the distribution subtract the mean from the value x *or* the mid-value X of that line.

(e) put the answers to (d) in the second new column
i.e. value − mean   or   mid-value − mean.
These are the deviations from the mean.

(f) for each line in the distribution, multiply the deviation in the second new column by the frequency of that deviation.

(g) *ignoring all minus signs* put the answers to (f) in the third new column,
i.e. frequency × (value−mean) or frequency × (mid-value − mean).

(h) add up the third new column. This gives the total of the deviations from the mean.

(i) add up the frequency column. This gives the number of values in the data.

(j) Divide the answer to (h) by the answer to (i).

(k) write down the answer to (j) as the mean deviation from the mean.

Example: Find the mean and the mean deviation from the mean of the frequency distribution below.

Value	1	2	3	4	5	6	7	8	9	10
Freq.	1	2	4	6	5	6	3	1	1	1

Rewriting the frequency distribution gives:

x Value	f Freq.	fx Freq. × Val.	$\bar{x}$ Mean	$x-\bar{x}$ Val. − Mean	$f(x-\bar{x})$ Freq. × (Val. − Mean)
1	1	1	5	−4	4
2	2	4	5	−3	6
3	4	12	5	−2	8
4	6	24	5	−1	6 ←minus
5	5	25	5	0	0 signs
6	6	36	5	1	6 ignored
7	3	21	5	2	6
8	1	8	5	3	3
9	1	9	5	4	4
10	1	10	5	5	5

$\Sigma f = 30$ $\Sigma fx = 150$ $\Sigma f(x-\bar{x}) = 48$

Number of   Total   Total Deviations

Values   Data

For the mean:

Total of Data = Σfx = 150

Number of values = Σf = 30

$$30 \overline{)150} \quad 5$$

Mean = 5

For the mean deviation from the mean:

Total of deviations = $\Sigma f(x-\bar{x})$ = 48

Number of values = Σf = 30

$$30 \overline{)48} \quad 1.6$$

Mean deviation from the mean = 1.6

The frequency distribution has a mean of 5, with a mean deviation from the mean of 1.6.

Example: Find the mean and mean deviation from the mean of the following distribution.

Class	1–5	6–10	11–15	16–20	21–25	26–30	31–35	36–40	41–45	46–50
Freq	1	3	5	6	10	14	4	3	2	2

Rewriting this frequency distribution gives:

Class	f Freq.	X Mid. Val.	fX Freq.×M.V.	$\bar{X}$ Mean	$X-\bar{X}$ M.V.−Mean	$f(X-\bar{X})$ Freq.×(M.V.−Mean)
1–5	1	3	3	25	−22	22
6–10	3	8	24	25	−17	51
11–15	5	13	65	25	−12	60
16–20	6	18	108	25	−7	42 ←minus
21–25	10	23	230	25	−2	20 signs
26–30	14	28	392	25	3	42 ignored
31–35	4	33	132	25	8	32
36–40	3	38	114	25	13	39
41–45	2	43	86	25	18	36
46–50	2	48	96	25	23	46

$\Sigma f=50$ $\Sigma fX=1250$ $\Sigma f(X-\bar{X})=390$

Number Values   Total Data   Total Deviations

For the mean: For the mean deviation from the mean:

Total of data = ΣfX = 1250   Total of deviations = $\Sigma f(X-\bar{X})$ = 390

Number of values = Σf =50   Number of values = Σf = 50

$$\begin{array}{r} 25 \\ \hline 50\overline{)1250} \end{array} \qquad\qquad \begin{array}{r} 7.8 \\ \hline 50\overline{)390} \end{array}$$

Mean = 25   Mean deviation from the mean = 7.8

The frequency distribution has a mean of 25, with a mean deviation from the mean of 7.8.

Neither the mean, nor the deviations from the mean, need be whole numbers as they were in the two previous examples. The table below demonstrates this.

Class	f	X	fX	$\bar{X}$	$X-\bar{X}$	$f(X-\bar{X})$
64–67	1	65.5	65.5	72.7	−7.2	7.2
68–71	4	69.5	278.0	72.7	−3.2	12.8
72–75	7	73.5	514.5	72.7	0.8	5.6 ←minus signs
76–79	3	77.5	232.5	72.7	4.8	14.4 ignored

$\Sigma f = 15$ $\Sigma fX = 1090.5$ $\Sigma f(X-\bar{X}) = 40.0$

For mean $\Sigma fX = 1090.5$ For mean deviation $\Sigma f(X-\bar{X}) = 40.0$

           $\Sigma f = 15$ $\Sigma f = 15$

$$\begin{array}{r} 72.7 \\ \hline 15\overline{)1090.5} \end{array} \qquad\qquad \begin{array}{r} 2.6\dot{6} \\ \hline 15\overline{)40.0} \end{array}$$

Mean = 72.7   Mean deviation = 2.7 to 1 decimal place

The distribution has a mean of 72.7, with a mean deviation from the mean of 2.7.

The mean deviation makes use of all the values in the data and is a better measure of dispersion than either the range or the semi-interquartile range, both of which use only a small number of the values.

As the minus signs are ignored in the calculation of the mean deviation, it cannot be developed further and so is of limited use.

Exercise 5a:

1. Find the mean deviation from the mean of the following sets of numbers.
   (a) 2   4   7   5   10   8   13
   (b) 0   3   5   9   6   16   10
   (c) 10   13   15   19   16   26   20
   (d) 2   6   12   14   16   18   21   23
   (e) 22   26   32   34   36   38   41   43

2. Find the mean deviation from the mean of the following frequency distribution.

x	0	1	2	3	4	5	6	7	8	9
f	2	2	3	5	7	4	3	2	1	1

3. Find the mean deviation from the mean of the following frequency distribution.

x	2	4	6	8	10	12	14	16	18	20
f	3	2	3	4	14	9	6	5	3	1

4. Find the mean deviation from the mean of the following frequency distribution.

x	1	2	3	4	5	6	7	8	9	10
f	1	1	3	5	7	15	8	5	3	2

5. Find the mean deviation from the mean of the following frequency distribution.

x	10	11	12	13	14	15	16	17	18	19	20
f	1	6	4	5	14	20	16	16	9	5	4

6. Find the mean deviation from the mean of the following frequency distribution.

Class interval	0– 4	5– 9	10– 14	15– 19	20– 24	25– 29	30– 34	35– 39	40– 44	45– 49	50– 54
f	1	4	5	3	8	3	5	3	2	5	1

7. Find the mean deviation from the mean of the following frequency distribution.

Class interval	0– 8	9– 17	18– 26	27– 35	36– 44	45– 53	54– 62	63– 71	72– 80
f	1	2	3	4	7	12	9	8	4

8. Find the mean deviation from the mean of the following frequency distribution.

Class interval	20– 22	23– 25	26– 28	29– 31	32– 34	35– 37	38– 40
f	2	3	11	15	18	8	3

9. Find the mean deviation from the mean of the following frequency distribution.

Class interval	1– 10	11– 20	21– 30	31– 40	41– 50	51– 60	61– 70	71– 80	81– 90	91– 100
f	1	3	5	7	13	9	7	3	2	0

10. Find the mean deviation from the mean of the following frequency distribution.

Class interval	0– 9	10– 19	20– 29	30– 39	40– 49	50– 59	60– 69	70– 79	80– 89	90– 99
f	2	1	5	5	6	10	3	4	2	2

11. Use the data you collected about your class to find the mean deviation from the mean of the:

(a) weights,   (c) waist sizes,   (e) chest sizes,
(b) heights,   (d) hip sizes,

of the pupils in your class.

12. Repeat question 11 considering only the boys in your class.

13. Repeat question 11 considering only the girls in your class.

Standard Deviation

The *square* of a number is the result when that number is multiplied by itself.
E.g. the square of 3 is 9 since $3 \times 3 = 9$.
  the square of -4 is 16 since $(-4) \times (-4) = +16$.
The square of a number Y can be written as Y^2.
E.g. the square of $6 = 6^2$ and $6^2 = 6 \times 6 = 36$.
  the square of $-8 = (-8)^2$ and $(-8)^2 = (-8) \times (-8) = +64$.
It should be noted from the above examples that the square of any number is positive. This is due to the fact that a plus times a plus gives a plus and a minus times a minus also gives a plus.
The *square root* of a number B is the number A which when multiplied by itself will give B.
E.g. the square root of 81 is 9 since $9 \times 9 = 81$ but also
  $(-9) \times (-9) = 81$ so the square root of $81 = -9$ or $+9$.
There are always two square roots of a number, one which is positive and another which is negative.
The square root of B is written as $\sqrt{B}$.
E.g. the square root of 100 is written as $\sqrt{100}$ and $\sqrt{100} = +10$ or -10.
  the square root of 144 is written as $\sqrt{144}$ and $\sqrt{144} = +12$ or -12.
The mean deviation from the mean is a useful measure of dispersion, but because the minus signs of the deviations are ignored, its usefulness is limited in more advanced work.
If the deviations from the mean are squared, all of the squares obtained will be positive, even those of negative deviations. This is a method of dealing with the minus signs which *does not* ignore the minus signs.
The mean deviation from the mean was a measure of dispersion which was worked out by finding the mean of the deviations from the mean.

The *variance* is another measure of dispersion, which is worked out by finding the mean of the *squares* of the *deviations* from the mean.

The variance is a perfectly acceptable measure of dispersion, except for the fact that it does not have the same units as the values in the original array of data.
The units of the variance will be the units of the original array, squared.
E.g. if the units of the original array are cm then the variances will be in cm^2.
When working out the variance of a set of data the *deviations from the mean* must be used. The deviations from the median or mode *cannot* be used.

Example: Find the variance of the array below. All of the values in the array are measured in seconds.

| 1 | 2 | 2 | 2 | 3 | 3 | 5 | 6 | 6 | 10 |

The mean of the array is 4 secs.
The deviations from the mean are:

$$-3 \quad -2 \quad -2 \quad -2 \quad -1 \quad -1 \quad 1 \quad 2 \quad 2 \quad 6$$

The squares of the deviations are:

$$(-3)^2 \;\; (-2)^2 \;\; (-2)^2 \;\; (-2)^2 \;\; (-1)^2 \;\; (-1)^2 \quad 1^2 \quad 2^2 \quad 2^2 \quad 6^2$$
$$9 \qquad 4 \qquad 4 \qquad 4 \qquad 1 \qquad 1 \qquad 1 \qquad 4 \qquad 4 \qquad 36$$

The mean of the squares of the deviations is $68 \div 10 = 6.8$.
The variance of the array is $6.8\ \text{s}^2$.
The original array had units of seconds but the variance has units of s^2.
The problem of units is solved if we take the square root of the variance, since the units of the square root of the variance will then be the same as those of the original array.
The square root of the variance is called the *standard deviation* and it is the most important measure of dispersion.
The standard deviation is the *positive* square root of the variance.
E.g. in the example above, standard deviation $= \sqrt{6.8\ \text{s}^2} = 2.61\text{s}$ (from tables).
As standard deviation is the square root of the variance, and variance is the mean of the squares of the deviations from the mean, we can define standard deviation as follows:
Standard deviation is the square root of the mean of the squares of the deviations from the mean.
This is often shortened to *root mean square deviation*.
For any set of data we can find the standard deviation by first working out the variance, and then taking the square root of the variance.

Examples:

Find the standard deviation of
98 99 100 101 102

Mean $= 100$

Deviations from the mean are
$-2 \quad -1 \quad 0 \quad 1 \quad 2$

Squares of the deviations are
$(-2)^2 \;\; (-1)^2 \quad 0^2 \quad 1^2 \quad 2^2$
$\;\;4 \qquad 1 \qquad 0 \qquad 1 \qquad 4$

Mean of squares $= 10 \div 5 = 2$
Variance $= 2$
Standard deviation $= \sqrt{2}$
$= 1.4$ to one
decimal place.

Find the standard deviation of
0 50 100 150 200

Mean $= 100$

Deviations from the mean are
$-100 \quad -50 \quad 0 \quad 50 \quad 100$

Squares of the deviations are
$(-100)^2 \;\; (-50)^2 \;\; 0^2 \quad 50^2 \qquad 100^2$
$10,000 \quad 2,500 \quad 0 \quad 2,500 \quad 10,000$

Mean of squares $= 25,000 \div 5 = 5,000$
Variance $= 5,000$
Standard deviation $= \sqrt{5,000}$
$= 70.7$ to one
decimal place.

$$\text{Variance} = \frac{\text{Total of the squares of the deviations}}{\text{Number of values}}$$

Standard Deviation $= \sqrt{\text{Variance}}$ *Note:* positive root

$$= \sqrt{\frac{\text{Total of the squares of the deviations}}{\text{Number of values}}}$$

To find the standard deviation of a frequency distribution:

(*a–f*) proceed as for mean deviation described on pages 119–24.

(*g*)   as for mean deviation but *do not* ignore the signs.

(*h*)   make a fourth new column headed:

for un-grouped data	$f(x-\bar{x})^2$ freq.×deviation² freq.×(val.−mean)²	or	$f(X-\overline{X})^2$ freq.×deviation² freq.×(M.V.−mean)²	for grouped data.

(*i*)   for each line of the distribution, multiply the deviation in the second new column by the freq. × deviation of the third new column.

(*j*)   put the answers to (*i*) in the fourth new column, i.e. freq.×(val.−mean)² *or* freq.×(M.V.−mean)².

(*k*)   add up the fourth new column. This gives the total of the squares of the deviations from the mean.

(*l*)   add up the frequency column. This gives the number of values in the data.

(*m*) divide the answer to (*k*) by the answer to (*l*). This gives the variance.

(*n*)   find the square root of the answer to (*m*).

(*o*)   write down the answer to (*n*) as the standard deviation.

Example: Find the mean and standard deviation of the frequency distribution below.

Value	1	2	3	4	5	6	7	8	9	10
Freq.	1	2	4	6	5	6	3	1	1	1

Rewriting the frequency distribution gives:

x Value	f Freq.	fx Freq.×Val.	$\bar{x}$ Mean	$x-\bar{x}$ Val−Mean	$f(x-\bar{x})$ Freq.×(Val−mean)	$f(x-\bar{x})^2$ Freq.×(Val−Mean)²
1	1	1	5	−4	−4	16
2	2	4	5	−3	−6	18
3	4	12	5	−2	−8	16
4	6	24	5	−1	−6	6
5	5	25	5	0	0	0
6	6	36	5	1	6	6
7	3	21	5	2	6	12
8	1	8	5	3	3	9
9	1	9	5	4	4	16
10	1	10	5	5	5	25

(notes in the table: "minus signs not ignored" beside the $f(x-\bar{x})$ column; "minus times minus gives plus" beside the $f(x-\bar{x})^2$ column)

$\Sigma f = 30 \qquad \Sigma fx = 150 \qquad\qquad\qquad \Sigma f(x-\bar{x})^2 = 124$

Number of   Total   Total of squares of
Values   Data   deviations.

For the mean: For the standard deviation:
Total of Data=Σfx=150   Total of squares of deviations=
                                        $\Sigma f(x-\bar{x})^2$=124

Number of values=Σf=30   Number of values=Σf=30
$$\underset{30\overline{)150}}{5}$$ $$\underset{30\overline{)124}}{4.133}$$
Mean=5   Variance=4.133
                                 Standard deviation= $\sqrt{4.133}$=2.03 to
                                 2 dec. places

The frequency distribution has a mean of 5 and a standard deviation of 2.03.

Example: Find the mean and standard deviation of the frequency distribution below.

Class	1– 5	6– 10	11– 15	16– 20	21– 25	26– 30	31– 35	36– 40	41– 45	46– 50
Freq.	1	3	5	6	10	14	4	3	2	2

Rewriting the frequency distribution gives:

Class	f Freq.	X Mid. Val.	fX Freq. × M.V.	$\overline{X}$ Mean	$X-\overline{X}$ M.V.– Mean	$f(X-\overline{X})$ Freq.× (M.V.–Mean)	$f(X-\overline{X})^2$ Freq.× (M.V.–Mean)²
1–5	1	3	3	25	−22	−22	484
6–10	3	8	24	25	−17	−51	867
11–15	5	13	65	25	−12	−60	720
16–20	6	18	108	25	−7	−42	294
21–25	10	23	230	25	−2	−20	40
26–30	14	28	392	25	3	42	126
31–35	4	33	132	25	8	32	256
36–40	3	38	114	25	13	39	507
41–45	2	43	86	25	18	36	648
46–50	2	48	96	25	23	46	1058

$\Sigma f=50$　　$\Sigma fX=1250$　　　　　　　　$\Sigma f(X-\overline{X})^2=5000$

Number of values　Total Data　　　　　　Total of squares of deviations

For the mean:　　　　　　　For the standard deviation:

Total of Data$=\Sigma fX=1250$　　Total of squares of deviations$=\Sigma f(X-\overline{X})^2=$
　　　　　　　　　　　　　　　5000

Number of values$=\Sigma f=50$　　Number of values$=\Sigma f=50$

$$\frac{25}{50)\overline{1250}}$$　　　　　　　$$\frac{100}{50)\overline{5000}}$$

Mean$=25$　　　　　　　　Variance$=100$

　　　　　　　　　　　　Standard deviation$=\sqrt{100}=10$

The frequency distribution has a mean of 25 and a standard deviation of 10.

The following table shows the working for a question involving decimals.

Class	f	X	fX	$\overline{X}$	$X-\overline{X}$	$f(X-\overline{X})$	$f(X-\overline{X})^2$
64–67	1	65.5	65.5	72.7	−7.2	−7.2	51.84
68–71	4	69.5	278.0	72.7	−3.2	−12.8	40.96
72–75	7	73.5	514.5	72.7	0.8	5.6	4.48
77–79	3	77.5	232.5	72.7	4.8	14.4	69.12

$\Sigma f=15$　$\Sigma fX=1090.5$　　　　　　　$\Sigma f(X-\overline{X})^2=166.40$

For the mean:　　　　　　　For the standard deviation:

　　$\Sigma fX=1090.5$　　　　　　$\Sigma f(X-\overline{X})^2=166.40$

　　　$\Sigma f=15$　　　　　　　　$\Sigma f=15$

$$\frac{72.7}{15)\overline{1090.5}}$$　　　　　　$$\frac{11.093}{15)\overline{166.40}}$$

Mean$=72.7$　　　　　　　Variance$=11.09$

　　　　　　　　　　　Standard deviation$=\sqrt{11.09}=3.33$

Exercise 5b:
Repeat all questions of Exercise 5a but find the standard deviation in each question instead of the mean deviation.

Mean and Standard Deviation
The mean and standard deviation of data obtained (transformed) from other data.
When a new set of data is obtained from another set of data by adding (or subtracting) a constant value N from all of the values in the original set of data then:

> (a) the mean of the new set of data is equal to the mean of the original set of data plus (or minus) the constant value N.
> (b) the standard deviation of the new set of data is the *same* as the standard deviation of the original set of data.

Consider   0   3   4   7   11   as an original set of data
     and   20   23   24   27   31   as a new set of data
obtained by adding 20 to each of the values in the original set of data.
The mean of   0   3   4   7   11 is 5.
The standard deviation of   0   3   4   7   11 is 3.74 to 2 D.P.
Using the above statements (a) and (b) we have:
The mean of   20   23   24   27   31 is 5 + 20=25.
The standard deviation of   20   23   24   27   31 is 3.74 to 2 D.P.

When a new set of data is obtained from another set of data by multiplying (or dividing) each of the values in the original set of data by a constant value M then:

> (c) the mean of the new set of data is equal to the mean of the original set of data multiplied (or divided) by the constant value M.
> (d) the standard deviation of the new set of data is equal to the standard deviation of the original set of data multiplied (or divided) by the constant value M.

Consider   1   4   6   7   2   as an original set of data
     and   10   40   60   70   20   as a new set of data
obtained by multiplying each of the values in the original set of values by 10.
The mean of   1   4   6   7   2 is 4
The standard deviation of   1   4   6   7   2 is 2.28 to 2 D.P.
Using the above statements (c) and (d) we have:
The mean of   10   40   60   70   20 is $4 \times 10 = 40$
The standard deviation of   10   40   60   70   20 is $2.28 \times 10 = 22.8$

Percentiles

It is often useful to know the value in a set of data below which a given percentage of the values lie,
e.g. which value has 60% of the values below it.
The value with 60% of the values below it is called the 60th percentile.
The value with 45% of the values below it is called the 45th percentile.
The value with $N\%$ of the values below it is called the Nth percentile.
The *Nth percentile* of a set of data is the value in the data which has $N\%$ of the values below it.
The 25th percentile is the lower quartile, Q_1.
The 50th percentile is the median.
The 75th percentile is the upper quartile, Q_3.
The *deciles* of a set of data are the percentiles which are multiples of 10,
e.g. the 1st decile is the 10th percentile.
the 2nd decile is the 20th percentile.
the 8th decile is the 80th percentile.
To find the Nth percentile of a set of data:

(a) draw a cumulative frequency curve showing 'less than' (see pages 56–7).
(b) find the cumulative frequency on the vertical (up) axis which is the same as the highest cumulative frequency *plotted* on the curve.
(c) divide the answer to (b) by 100.
(d) multiply the answer to (c) by N.
(e) find the point on the vertical (up) axis which gives the same cumulative frequency as the answer to (d).
(f) draw a line from this point on the vertical (up) axis *across* to the curve.
(g) draw a line from where the line in (f) cuts the curve *down* to the horizontal (across) axis.
(h) read off the number where the line in (g) cuts the horizontal (across) axis.
(i) write down the answer to (h) as the Nth percentile.

Example: Find the 65th percentile of the cumulative frequency distribution below.

Marks <	10	20	30	40	50	60	70	80	90	100
Cum. Freq.	10	30	70	170	300	420	540	640	690	700

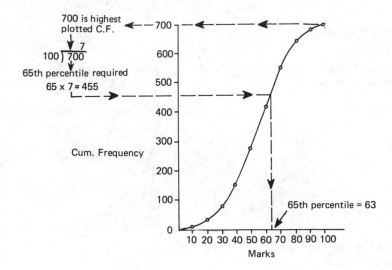

The difference between the 90th percentile and the 10th percentile is called the inter-percentile range.

The inter-percentile range is a measure of dispersion similar to the semi-interquartile range.

80% of the values lie inside the inter-percentile range as opposed to 50% lying between the quartiles, i.e. 30% more of the values of the distribution are used to find the inter-percentile range, than are used to find the semi-interquartile range.

The bottom 10% and the top 10% of the values do not affect the inter-percentile range, so it is not affected unduly by very high or very low values.

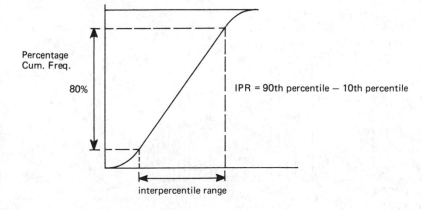

Frequency Curves

A large number of short lines can give the appearance of being a curve, e.g. curve stitching patterns.

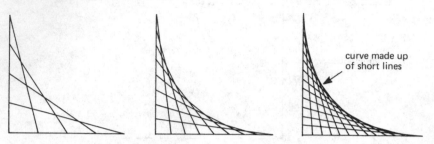

curve made up of short lines

As the number of lines increases and their lengths become shorter, the more they resemble a curve.

For a large population, data can be classed in a large number of small sized class intervals.

Intervals with small class sizes will have narrow bars on a histogram.

A histogram of data from a large population will have a large number of narrow bars.

Since frequency polygons are based on histograms, a frequency polygon of a large population will be made up of a large number of short lines.

A frequency polygon will have very short lines when:

(*a*) the amount of data is large.
(*b*) the class sizes are small.

A frequency polygon with a large number of very short lines will have a shape very close to a curve.

The larger the population the more a frequency polygon will approximate to a frequency curve.

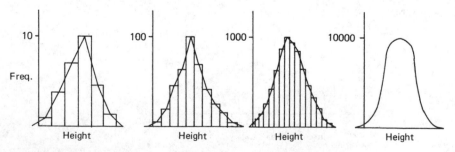

Frequency curves of statistical data should only be drawn when there is a large amount of data.

The general shape of a frequency curve is usually apparent from a frequency polygon.
Frequency curves can have many different shapes depending on the frequency distribution which they are representing.

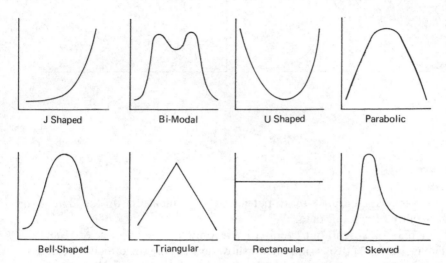

| J Shaped | Bi-Modal | U Shaped | Parabolic |

| Bell-Shaped | Triangular | Rectangular | Skewed |

The frequency distribution which occurs most in nature is the *normal curve*.
Distributions of heights of people, weights of people, the number of blades of grass per square metre, the number of peas in a pod, the way a step is worn away, exam marks, all have frequency curves which have the shape of the *normal curve*.
The normal curve is *bell-shaped*.

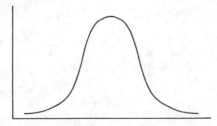

The normal curve is *symmetrical* (exactly the same shape on both sides) about a line drawn down through its highest point.
The areas under the curve on either side of this line are the same as each other.
On a normal curve the mode, median and mean all have the same value.
The mode, median and mean are given by the value on the horizontal (across) axis where the line drawn down from the highest point cuts the axis.

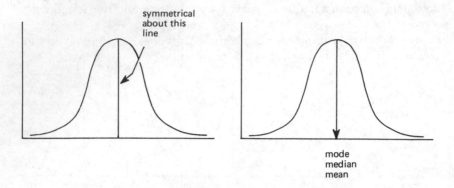

Skewness

The various measures of dispersion give a measure of the spread or variability of a set of data.

Data may be spread out evenly or unevenly.

The measures of dispersion do not show how evenly the data is spread out.

On a normal curve the data is spread out evenly on either side of the mean, i.e. the curve is symmetrical.

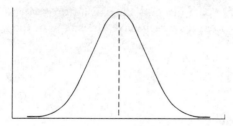

Frequency distributions whose curves should be *normal*, i.e. bell shaped, often turn out to be *non-symmetrical*.

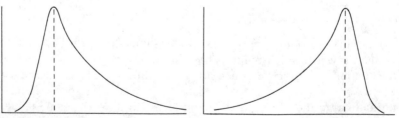

Distributions and curves which are non-symmetrical are called *skewed* distributions or curves.

There are two types of skewed curves:

(*a*) positive skewed

(*b*) negative skewed

(*a*) A *positive skewed* curve has a long tail on the right hand side, i.e. at the positive end of the horizontal (across) axis, and a short tail on the left hand side.

(*b*) A *negative skewed* curve has a long tail on the left hand side, i.e. at the negative end of the horizontal (across) axis, and a short tail on the right hand side.

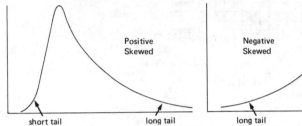 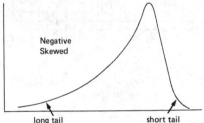

On a positive skewed curve the mean is bigger than the mode, i.e. mean > mode.

On a negative skewed curve the mean is smaller than the mode, i.e. mean < mode.

On either type of skewed curves the median is somewhere between the mode and the mean.

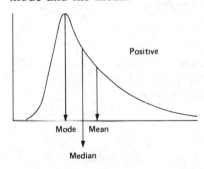 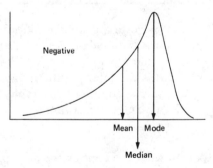

Provided a curve is not too skewed, the following approximate formula is true.

$$\text{Mode} - \text{median} = 2\,(\text{median} - \text{mean})$$

A measure of the skewness of a distribution can be calculated using Pearson's formula for the coefficient of skewness.

$$\text{Skew} = \frac{3\,(\text{mean} - \text{median})}{\text{standard deviation}}$$

Miscellaneous Exercises 5

1.

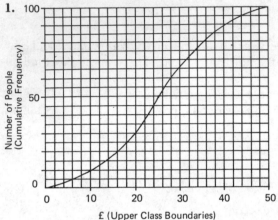

£ (Upper Class Boundaries)

Ogive of donations to charity
by 100 people

Class in £	Upper Class Boundary	Cumulative Frequency	Frequency
0.50– 9.50	10		
10.50–19.50	20		
20.50–29.50	30		
30.50–39.50	40		
40.50–49.50	50		

Use the ogive to answer the following questions.
  (i) What is the median donation?
 (ii) How many people gave less than £20 to the charity?
(iii) What is the interquartile range of donations?
(iv) Copy and complete the table to obtain the number of people
     who gave in each class, taking the cumulative frequency scores from the
     ogive.   NI77

2. Explain what is meant by the 40th percentile of a distribution. WM75

3. The following table gives the cumulative frequency for the distribution
of marks out of a maximum of 100 in an examination. No candidate
scored full marks.

Less than	10	20	30	40	50	60	70	80	90	100
Cumulative frequency	8	27	56	100	148	187	225	259	282	298

(a) How many candidates scored fewer than half marks?
(b) How many candidates scored 80 or more marks?
(c) Draw a cumulative frequency curve and use your curve to estimate
   (i) the median mark;
   (ii) the 30th percentile.   WM78

4. A firm has two different machines for bagging flour. In a random sample of one hundred bags from each machine the following frequency distributions were obtained:

Machine A

Weight (100 g)	14.7	14.8	14.9	15.0	15.1	15.2	15.3
Frequency	10	10	15	25	18	12	10

Machine B

Weight (100 g)	14.7	14.8	14.9	15.0	15.1	15.2	15.3
Frequency	5	7	18	40	20	5	5

   (i) calculate the range and mean for each distribution.
   (ii) plot a histogram and hence a frequency polygon for each distribution on the same axes.
   (iii) what is the mode for each distribution.
   (iv) calculate the interquartile range for each distribution.
   (v) comment on the efficiency of the two machines.

Note: In the above question the machine is designed to produce bags of 1,500 g.   EM78

5. The table below shows the frequency distribution of marks awarded to 1000 candidates in an examination.

Class of marks	1–10	11–20	21–30	31–40	41–50	51–60	61–70	71–80	81–90	91–100
Frequency of marks	18	39	93	127	203	210	171	88	31	20

(a) Construct a table showing the cumulative frequency distribution and draw the cumulative frequency curve (i.e. the ogive).
(b) Estimate from your graph:
   (i) the median mark,
   (ii) the mark at the 70th percentile,
   (iii) the number of candidates who got less than 35 marks,
   (iv) the percentile for a mark of 45.

(c) Briefly explain how you obtained *each* result. Check, by calculation and without reference to your graph, the accuracy of your four estimates. Comment on any assumptions you make.   M77

6. The table below gives the frequency distribution of marks obtained by 500 candidates in a statistics examination.

Mark	0–9	10–19	20–29	30–39	40–49	50–59	60–69	70–79	80–89	90–99
Frequency of marks	5	10	45	65	105	120	75	40	25	10

Construct a table showing the cumulative frequency distribution and draw the cumulative frequency curve (i.e. the ogive). Estimate from your graph:

(a) the median mark,
(b) the mark at the 90th percentile,
(c) the number of candidates who got less than 25 marks,
(d) the percentile for a mark of 40.

Briefly explain how you obtained *each* result. Check, by calculation and without reference to your graph, the accuracy of your four estimates. Comment on any assumptions you make.   M76

7. An enquiry was carried out into the number of apprentices employed in factories. Information was obtained from 600 factories and the results are shown in the table below.

No. of apprentices	0–14	15–29	30–44	45–59	60–74	75–89	90–104	105–119
No. of factories	22	46	86	108	147	99	60	32

Draw the cumulative frequency curve for this distribution.
From your curve find:
  (i) the median
  (ii) the 30th percentile
  (iii) the number of factories with more than 50 apprentices.   WM76

8. In a biological survey the weights of each of 200 plums (to the nearest gram) were as shown in the following table:

Weight (g)	30	31	32	33	34	35	36	37	38
No. of plums	2	5	18	44	61	45	20	4	1

(i) Compile a cumulative frequency table.
(ii) Draw a cumulative frequency curve.
(iii) Find the median.
(iv) Find the interquartile range.
(v) Find the sixtieth percentile.
(vi) Find the percentage of plums with a weight less than 32 g.   EM78

9. The following table gives the distribution of weekly incomes of a group of men

Income in £	20–	30–	40–	50–	60–	70–	80–	90–100
Frequency	15	42	65	92	75	67	37	7

(*a*) Copy and complete the following cumulative frequency table

Less than	20	30	40	50	60	70	80	90	100
Cumulative Frequency	0		57			289		393	

(*b*) Draw a cumulative frequency curve.
(*c*) Use your curve to estimate
 (i) the number of men who earn between £45 and £75 a week;
 (ii) the 80th percentile of the distribution;
 (iii) the interquartile range of the distribution.   WM76

10. The diagram represents the frequency distribution curve of the weekly wages of people in a factory. Draw, and label, vertical lines on the diagram to represent:

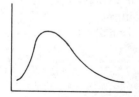

(*a*) the mode,
(*b*) the mean.

          WM77

11. (*a*) Draw a sketch-graph of:
 (i) a normal distribution curve,
 (ii) a positively skewed distribution curve.
 In the case of *each* of (i) and (ii) indicate on your diagram the estimated position of the mean of the distribution.

(*b*) Fifteen pupils are asked to estimate the length (to the nearest 2 cm) of their teacher's table. The estimates, arranged in descending order, were as follows:

| 148 | 146 | 144 | 142 | 140 | 140 | 138 | 138 | 138 | 136 | 136 | 134 | 132 | 132 | 128 |

Find the median estimate and the lower and upper quartiles. Also calculate the semi-interquartile range.   M76

12. (*a*) Draw a sketch graph of:
   (i) a normal distribution
   (ii) a positively skewed distribution.
   On each sketch graph mark, and label, the positions of the mean and the median.
   Give one simple example of the kind of data that you would expect to produce each of the two kinds of distribution.

(*b*) The weights, to the nearest kilogram, of 15 people in a keep-fit class are given below.

| 86 | 84 | 82 | 81 | 80 | 80 | 79 | 77 | 77 | 76 | 76 | 75 | 74 | 72 | 70 |

For these weights find (giving a brief explanation of how you obtained each answer):
   (i) the upper quartile weight
   (ii) the lower quartile weight
   (iii) the semi-interquartile range.   M78

13. (*a*) Draw a sketch-graph of:
   (i) a normal distribution
   (ii) a positively skewed distribution
   (iii) a negatively skewed distribution.
   Give one simple example of the kind of data that you would expect to produce each of the above kinds of distribution.

(*b*) The weights, to the nearest kilogram, of eleven pupils are given below.

| 63 | 62 | 61 | 60 | 60 | 59 | 57 | 56 | 55 | 52 | 50 |

For the above weights, find (giving a brief explanation of how you obtained each answer):
   (i) the upper quartile weight
   (ii) the lower quartile weight
   (iii) the semi-interquartile range.   M77

14.

Test Scores	Upper Class Limits	F Frequency	Cumulative Frequency
50.5–59.5	60	1	
60.5–69.5		3	
70.5–79.5		12	
80.5–89.5		23	
90.5–99.5		25	
100.5–109.5		22	
110.5–119.5		10	
120.5–129.5		3	
130.5–139.5		1	
		$\Sigma F = 100$	

(i) Copy and complete the table above onto your answer page.
(ii) Draw an ogive using the data.
   (Scales: x-axis – upper class limits – 2 cm represents 10 units
   y-axis – cumulative frequency – 2 cm represents 10 units.)
   From it determine:
(iii) the median
(iv) the quartiles
(v) the interquartile range of the scores.
(vi) If a frequency distribution curve were drawn its shape would be familiar. What is the name of the curve? NI78

15. Find the standard deviation of:
   1   3   4   5   7 WM75

16. Find the standard deviation of:
   2   4   5   6   8 WM76

17. Use your answer to question 16 to write down the standard deviation of:
   20   40   50   60   80 WM76

18. Use your answer to question 17 to write down the standard deviation of:
   25   45   55   65   85 WM76

19. Find the standard deviation of:
   2   6   8   10   14 WM78

20. Use your answer to question 19 to write down the standard deviation of:
   10   30   40   50   70 WM78

21. Use your answer to question 20 to write down the standard deviation of:

12 32 42 52 72   WM78

22. Find the standard deviation of:

1 3 5 9 12   WM77

23. Use the answer to question 22 to write down the standard deviation of:

11 13 15 19 22   WM77

24. Use the answer to question 23 to write down the standard deviation of:

55 65 75 95 110   WM77

25. The following numbers represent a batsman's scores in six cricket matches.

28 4 51 6 16 27

(*a*) Find the range of these scores.
(*b*) Calculate the standard deviation of these scores.   WM78

26. (*a*) The numbers 10, 10, 10, 10, 11 appear to be closely grouped. Determine their standard deviation and indicate whether it backs up the initial statement or not.

$$\text{S.D.} = \sqrt{\frac{\Sigma(x - \bar{x})^2}{n}}$$

(*b*) Determine the mean, median and mode of the following data:

Age of People in Sample in Years and Months

16–0	16–3	15–11	16–0	15–11
15–9	15–11	16–2	15–11	16–1
16–5	16–3	16–2	16–8	16–4
16–4	16–1	16–5	15–8	15–5

NI78

27. The marks of 16 pupils in a statistics examination were as follows.

50	70	52	47	38	64	27	48	20	85	55	42	77	49	33	59

Calculate the mean, variance and standard deviation of these marks, expressing your answer, where necessary, to a degree of accuracy which is appropriate for these given data.   M77

28. A group of 12 children were asked to keep a diary recording all the television programmes they watched during a particular week. At the end of the week the total viewing time for each child was worked out, with the following results.

Viewing times during 1 week, in hours, for 12 children:
16   10½   17½   5   12½   11   8   14   21½   16   0   12
For these times calculate: (i) the arithmetic mean,
(ii) the median,
(iii) the standard deviation.   WM75

29. The temperatures, in degrees Centigrade (°C) to the nearest degree, at noon on the 30 days in June 1977 in a certain holiday town, were as follows.

Temperature (°C)	16	17	18	19	20	21	22
Number of days	3	4	5	5	7	3	3

Calculate:

(*a*) the mean temperature
(*b*) the standard deviation of the temperature.   M78

30. A group of pupils kept a record of the number of times that they visited a cinema during each of 25 consecutive weeks. The figures below give the *total* number of attendances that these pupils made in each of these 25 weeks.

30	36	27	30	39	29	35	31	34	27
31	34	34	38	31	34	24	31	34	35
24	34	36	31	31					

Calculate the mean, variance and standard deviation of these numbers of attendances.   M76

31. The number of half-day absences for a class of 30 pupils in one term were as follows:

Half-day absences	3	4	5	6	7	8	9
Frequency	2	2	6	9	7	3	1

Calculate the mean and the standard deviation of this distribution. L78

32. (*a*)

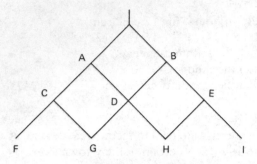

Tree diagram to show number
of routes in a binary system

The numbers of routes leading to *A* and *B* are   1   1
The numbers of routes leading to *C*, *D*, *E* are   1   2   1
The numbers of routes leading to *F*, *G*, *H*, *I* are   1   3   3   1
Determine the next 3 rows of this pattern (*Pascal's Triangle*).

(*b*) Find the standard deviation of the following tabulated data using the method and formulae given below.

Score	F Frequency	x Class Mid Value	Fx	$x - \bar{x}$	$(x - \bar{x})^2$	$F(x - \bar{x})^2$
5–9	1					
10–14	4					
15–19	11					
20–24	7					
25–29	2					
	$\Sigma F =$		$\Sigma Fx =$			$\Sigma F(x - \bar{x})^2$

$$\text{Mean} = \bar{x} = \frac{\Sigma Fx}{\Sigma F} \qquad \text{Standard deviation} = S = \sqrt{\frac{\Sigma F(x - \bar{x})^2}{\Sigma F}}$$

where ΣF means the sum of the frequencies.   NI78

33. (*a*) (i) State the mode and calculate the (ii) mean and (iii) median of the following data.

Weights of Twelve People (in Kg)

45	42	48	45
62	36	47	73
45	54	61	42

(*b*) Copy and complete the table and use it and the formulae below to obtain the standard deviation of the scores.

Score	F Frequency	X Class Mid-Value	FX	$X - \bar{X}$	$(X-\bar{X})^2$	$F(X-\bar{X})^2$
0–2	20					
3–5	20					
6–8	20					
9–11	20					
12–14	20					
	$\Sigma F = 100$		$\Sigma FX =$			$\Sigma F(X-\bar{X})^2 =$

$$\text{Mean} = X = \frac{\Sigma FX}{\Sigma F} \qquad \text{Standard Deviation} = S = \sqrt{\frac{\Sigma F(X-\bar{X})^2}{\Sigma F}}$$

where ΣF means the sum of the frequencies. NI77

34. Planks in a load of timber had an arithmetic mean length of 5 m and variance 1.21 m². How long would a plank be which was two standard deviations above the mean? EA78

35. Given that the standard deviation of the integers from 1 to 10 inclusive is 2.87, find the standard deviation of the integers from 11 to 20 inclusive. EA78

36. Using the information given in question no. 35 find the standard deviation of the first ten even integers. EA78

6 PROBABILITY 1

Probability or Chance

Throughout history people have gambled, i.e. taken a risk in the hope of some gain.

People have gambled for money on such things as horse races, cock fights, card games, movements of shares on the stock market, football pools and even when men would land on the moon.

People have gambled for power by starting revolutions and wars.

Some people have gambled with their own lives by playing Russian Roulette.

In any gambling situation the gambler works out his chance of winning either
  (a) on available evidence
or (b) by reasoning.

He may express his chance of winning by a statement such as 'a dead cert', 'odds on', 'evens', 'not much chance', or 'no chance'.

The more certain he is of his chances of winning, the more likely he is to gamble.

The chances or probabilities of events happening are of interest to many people who have no interest whatsoever in gambling.

E.g. a person may want to know the probability of:

  (a) a light bulb lasting more than 1000 hours,
  (b) dying of a heart attack before the age of 45,
  (c) being successful in an examination.

The probabilities of most events can be calculated with mathematical precision.

In mathematics vague descriptions such as 'odds on' are not acceptable as statements of the probabilities of events happening.

A mathematical probability is expressed as a fraction or decimal between 0 and 1 inclusive,
e.g. $\frac{1}{4}$ or 0.25, $\frac{3}{5}$ or 0.6.

A probability of 1 indicates that an event is certain to happen.

E.g. the probability that a brick thrown in the air will come down again is stated as 1.

A probability of 0 indicates that an event cannot possibly happen.

E.g. the probability that a pig will fly unaided is stated as 0.

The probability that an event will happen can be worked out by:

  (a) experiment   (b) reasoning.

Probability by Experiment

An experiment is a procedure designed to find the results which follow from a particular course of action,

e.g. what happens when a coin is tossed,

what happens when a pack of cards is cut,

what happens when a spring is loaded with a weight.

An experiment consists of a series of tests or trials,

e.g. a series of tosses of a coin,

a series of cuts of a pack of cards,

a series of loadings of a spring.

At each trial during an experiment a number of outcomes are possible,

e.g. when a coin is tossed the possible outcomes are heads or tails.

when a drawing pin is tossed it can land either point up $\perp$ or point down $\nwarrow$.

when a die is thrown it may land showing the possible outcomes: 1, 2, 3, 4, 5 or 6.

An *event* is an outcome in which we are interested.

If an experiment is conducted the result of each trial can be entered on a tally chart against the correct outcome. This tally chart will give the overall results obtained during the experiment.

From a tally chart showing the results of an experiment we can find the probability that an event will happen.

If we wish to know the probability or chance of an event happening we can find it using the following formula,

Probability of an event happening =

$$\frac{\text{Number of times that event occurred during experiment}}{\text{Total number of trials observed during experiment}}$$

provided all trials are conducted under exactly similar conditions.

The probability of an event worked out by experiment is an *estimate* of the actual probability of that event.

E.g. the results obtained when a drawing pin was thrown 50 times are shown in the table below.

Point up $\perp$	*Point Down* $\nwarrow$	*Total Throws*
35	15	50

From the table we can find an estimate of the probability that a drawing pin will land point up from

Probability of pin landing point up =

$$\frac{\text{Number of times pin landed point up}}{\text{Total number of throws}}$$

$$= \tfrac{35}{50} = \tfrac{7}{10} = 0.7$$

Similarly from the table we can find an estimate of the probability that a drawing pin will land point down from

Probability of pin landing point down =

$$\frac{\text{Number of times pin landed point down}}{\text{Total number of throws}}$$

$$=\tfrac{15}{50} = \tfrac{3}{10} = 0.3$$

Statements such as:

the probability of drawing an ace,

the probability of tossing a head,

the probability of an event happening,

are usually shortened by writing P instead of 'the probability of', and writing the event under consideration in brackets after the P.

The shorthand notation gives P(Ace), P(Head) and P(Event) for the above statements.

In our example for drawing pins we would have

$$P(\perp) = 0.7 \text{ and } P(\wedge) = 0.3$$
and
$$P(\perp) + P(\wedge) = 0.7 + 0.3$$
$$= 1.$$

The two probabilities have a *total* of 1.

This is not surprising as it indicates mathematically the phrase 'what goes up must come down'.

In general the *sum* of the probabilities of all of the events which follow from a given course of action is 1.

E.g. if a course of action can produce the outcomes A, B, C, D, E then

$$P(A) + P(B) + P(C) + P(D) + P(E) = 1.$$

For example, when a coin is tossed

$$P(\text{Head}) + P(\text{Tail}) = 1$$
or
$$P(\text{Head}) + P(\text{not Head}) = 1.$$

When a normal six-sided die is tossed

$$P(1) + P(2) + P(3) + P(4) + P(5) + P(6) = 1$$
or
$$P(1) + P(\text{not } 1) = 1$$
where
$$P(\text{not } 1) = P(2) + P(3) + P(4) + P(5) + P(6).$$

If an experiment consists of a *large* number of trials, then the probabilities of events obtained from the results of the experiment will be good estimates of the actual probabilities of those events.

The *larger* the number of trials the better the estimates will become.

The progressive totals and probabilities obtained when a drawing pin was thrown 4500 times are shown in the table below.

Freq ⊥	Freq ⅄	Total Throws	P(⊥)	P(⅄)
69	31	100	0.690	0.310
134	66	200	0.670	0.330
196	104	300	0.653	0.347
266	134	400	0.665	0.335
333	167	500	0.667	0.333
655	345	1000	0.655	0.345
994	506	1500	0.663	0.337
1308	692	2000	0.654	0.346
1642	858	2500	0.657	0.343
1969	1031	3000	0.656	0.344
2293	1207	3500	0.655	0.345
2622	1378	4000	0.656	0.344
2948	1552	4500	0.655	0.345

From the above table it can be seen that as the total number of throws increases, the P(⊥) is settling down to a figure of 0.66 to 2 decimal places and P(⅄) to 0.34.

Note: At each stage P(⊥) + P(⅄) = 1.

Some probabilities can only be worked out by experimentation, e.g.

How a drawing pin falls when tossed.

How a fork falls when tossed.

How long a light bulb lasts.

To obtain good estimates of the probabilities of events by experimentation, the experiment should be continued until the number of trials has become large enough for the probabilities to settle down around a *constant* figure. The experiment with drawing pins gave P(⊥) = 0.66 or 66/100.

A probability of $\frac{66}{100}$ indicates that *on average* the drawing pin landed point up 66 times in every 100 throws.

Similarly if the probability of an event is 1/6 this probability indicates that on average the event will occur once in every six trials.

Probability by Reasoning

Many objects show *symmetry* in their construction.

E.g. a normal die has 6 faces each of exactly the same size and shape, a coin has 2 faces each of exactly the same size and shape.

When a symmetrical object, such as a die is tossed, it seems reasonable to assume that each face has an equally likely chance of showing uppermost.

Many objects are *congruent*, i.e. have exactly the same shape and size, e.g.

All marbles made by manufacturer A are congruent to each other.
All dominoes made by manufacturer B are congruent to each other.
All raffle tickets made by manufacturer C are congruent to each other.
All pencils made by manufacturer D are congruent to each other.
All playing cards made by manufacturer E are congruent to each other.

When a set of congruent items is placed in a bag, it seems reasonable to assume that each item has an equally likely chance of being drawn from the bag at the first pick.

When it is apparent that the outcomes of a particular course of action are all equally likely, it is possible to work out the probabilities of events by *reasoning*, i.e. in theory without the need for experimentation.

If it is possible to *reason* out the possible outcomes which follow from a particular course of action, we can find the probability that an event will occur from the formula

$$P(\text{Event}) = \frac{\text{Number of outcomes } \textit{favourable} \text{ to the event}}{\textit{Total} \text{ number of outcomes possible}}$$

provided all outcomes are equally likely.

If the number of outcomes favourable to the event $= N$ and total number of outcomes possible $= T$

$$P(\text{Event}) = \frac{N}{T}$$

Example: Find the probability of throwing a head with one toss of a coin.
Number of outcomes favourable to head $= N = 1$
Total number of outcomes possible $\quad = T = 2$.

$$P(\text{Head}) = \frac{N}{T} = \frac{1}{2}.$$

Example: Find the probability of drawing an ace from a pack of cards.
Number of outcomes favourable to ace $= N = 4$
Total number of outcomes possible $\quad = T = 52$

$$P(\text{Ace}) = \frac{N}{T} = \frac{4}{52} = \frac{1}{13}.$$

Example: Find the probability of throwing a normal six-sided die and obtaining a four with one throw of the die.
Number of outcomes favourable to 4 $= N = 1$
Total number of outcomes possible $\quad = T = 6$

$$P(4) = \frac{N}{T} = \frac{1}{6}.$$

Example: A bag contains 15 red marbles, 20 green marbles and 45 blue marbles. What is the probability of pulling out a green marble at the first attempt?

Number of outcomes favourable to green $= N = 20$
Total number of outcomes possible $\quad = T = 80$

$$P(\text{Green}) = \frac{N}{T} = \frac{20}{80} = \frac{1}{4}.$$

Probability of an event not happening

Since an event A will either happen

or *not* happen

the sum of the probabilities $P(A)$ and $P(\text{not } A)$ must equal 1,

$$P(A) + P(\text{not } A) = 1$$

so $\qquad P(\text{not } A) = 1 - P(A).$

To find the probability of an event *not* happening:

(*a*) find the probability that the event will happen.
(*b*) subtract the answer to (*a*) from 1.

Example: Find the probability that when a normal six-sided die is thrown it will not show a 2 uppermost.

(*a*) To find the probability that the die will show 2 uppermost:
Number of outcomes favourable to $2 = N = 1$.
Total number of outcomes possible $\quad = T = 6$.

$$P(2) = \frac{N}{T} = \frac{1}{6}$$

(*b*) $\qquad P(\text{not } 2) = 1 - P(2)$
$$= 1 - \tfrac{1}{6} = \tfrac{5}{6}.$$

Example: If the letters of the word MISSISSIPPI are each written on a piece of paper and dropped into a bag, what is the probability of drawing a piece of paper out of the bag which does not have the letter I on it?

(*a*) Number of outcomes favourable to I $= N = 4$.
Total number of possible outcomes $\quad = T = 11$.

$$P(I) = \frac{N}{T} = \frac{4}{11}$$

(*b*) $P(\text{not } I) = 1 - P(I)$
$$= 1 - \tfrac{4}{11} = \tfrac{7}{11}.$$

Sample Spaces

A list or table of all of the possible outcomes which follow from a particular course of action is called a *sample space.*

E.g. the sample space when a coin is tossed is

Head Tail

The sample space when a normal six-sided die is tossed is

1 2 3 4 5 6

The probabilities of events can be worked out from sample spaces using the formula

$$P(\text{Event}) = \frac{\text{Number of outcomes in sample space favourable to event}}{\text{Total Number of possible outcomes in a sample space}}$$

Example: Find the probability that when a die is tossed once the die will show an even number uppermost.

Sample Space is 1 2 3 4 5 6.

Number of outcomes in sample space favourable to an even number = N = 3.

Total number of possible outcomes in sample space = 6.

$$P(\text{Even Number}) = \frac{N}{T} = \frac{3}{6} = \frac{1}{2}$$

Example: If counters numbered 21 to 60 inclusive are placed in a bag, what is the probability that a counter pulled out of the bag has a number on it which is:

(*a*) a multiple of 5?
(*b*) a multiple of 3?
(*c*) a multiple of both 3 and 5?
(*d*) a multiple of either 3 or 5?

The sample space is

㉑	22	23	㉔	[25]	26	㉗	28	29	[㉚]
31	32	㉝	34	[35]	㊱	37	38	㊴	[40]
41	㊷	43	44	[㊺]	46	47	㊽	49	[50]
(51)	52	53	(54)	[55]	56	(57)	58	59	[(60)]

Note: A number is a multiple of a number Y if it is exactly divisible by Y.

The multiples of 3 in the sample space have been marked ○.

The multiples of 5 in the sample space have been marked □.

It follows that the multiples of both 3 and 5 are marked [○].

(*a*) Number of outcomes in sample space which are multiples of 5, i.e. marked $\square = N = 8$.
Total number of possible outcomes in sample space $= T = 40$.

$$P(\text{Multiple of } 5) = \frac{N}{T} = \frac{8}{40} = \frac{1}{5}$$

(*b*) Number of outcomes in sample space which are multiples of 3, i.e. marked $\bigcirc = N = 14$.
Total number of possible outcomes in sample space $= T = 40$.

$$P(\text{Multiple of } 3) = \frac{N}{T} = \frac{14}{40} = \frac{7}{20}$$

(*c*) Number of outcomes in sample space which are multiples of both 3 and 5, i.e. marked $\square = N = 3$.
Total number of possible outcomes in sample space $= T = 40$.

$$P(\text{Multiple of both 3 and 5}) = \frac{N}{T} = \frac{3}{40}$$

(*d*) Number of outcomes in sample space which are multiples of either 3 or 5, i.e. marked either $\bigcirc$ or $\square$ or $\square = N = 19$.
Total number of possible outcomes in sample space $= T = 40$.

$$P(\text{Multiple of either 3 or 5}) = \frac{N}{T} = \frac{19}{40}$$

Example: If counters numbered with the squares of the numbers from 1–15 inclusive are placed in a bag, what is the probability that a counter pulled from the bag has a number on it whose digits:

(*a*) total 9?
(*b*) total less than 9?
(*c*) total not less than 9?

The squares of the numbers 1–15 inclusive are
1 4 9 16 25 36 49 64 81 100 121 144 169 196 225.
The sample space of the totals of the digits on the counters is
1 4 9 7 7 9 13 10 9 1 4 9 16 16 9.

(*a*) Number of outcomes in sample space which are 9 $= N = 5$.
Total number of possible outcomes in sample space $= T = 15$.

$$P(\text{Total of } 9) = \frac{N}{T} = \frac{5}{15} = \frac{1}{3}$$

(*b*) Number of outcomes in sample space which are less than 9 $= N = 6$.
Total number of possible outcomes in sample space $= T = 15$.

$$P(\text{Total less than 9}) = \frac{N}{T} = \frac{6}{15} = \frac{2}{5}$$

(c) P(Total not less than 9) = 1 − P(Total less than 9)
$$= 1 - \tfrac{2}{5} = \tfrac{3}{5}.$$

Combinations of Happenings

Sample spaces are very useful when the probabilities of events which result from the combination of the outcomes of two or more separate happenings are required.

E.g. what is the probability of throwing a coin and a die together and obtaining a head on the coin and a six on the die?

The sample spaces for each of the separate happenings can be combined together into a *composite* sample space from which probabilities can be worked out as previously described.

A composite sample space for 2 or more happenings is a table which shows all of the possible outcomes when the happenings are combined together.

To construct a composite sample space for 2 happenings:

(a) write out the sample space for one of the happenings across the page.
(b) write out the sample space for the other happening down the page.
(c) fill in the body of the table with the possible outcomes obtained by combining the two sample spaces from (a) and (b).

	Sample space 1
Sample *space 2*	Body of Table

Composite sample space.

Example: Find the probability that when a coin and a normal six sided die are thrown together the result will be:

(a) a head and a six.
(b) a tail and an odd number.

The sample space for the die is 1 2 3 4 5 6.
The sample space for the coin is Head Tail.
The composite sample space is given by:

		Sample space 1					
		1	*2*	*3*	*4*	*5*	*6*
Sample	*Head*	H1	H2	H3	H4	H5	H6
space 2	*Tail*	T1	T2	T3	T4	T5	T6

Composite sample space.

(*a*) Number of outcomes in composite sample space which are H6 $=N$
  $= 1$.
  Total number of possible outcomes in composite sample space $= T$
  $= 12$.

$$P(\text{Head and six}) = \frac{N}{T} = \frac{1}{12}$$

(*b*) Number of outcomes in composite sample space which are T and
  Odd $= N = 3$.
  Total number of possible outcomes in composite sample space $= T$
  $= 12$.

$$P(\text{Tail and odd number}) = \frac{N}{T} = \frac{3}{12} = \frac{1}{4}$$

Example: If 2 dice are thrown at the same time, what is the probability that
the total of the numbers uppermost on the 2 dice is 10?
The sample spaces for both dice are 1   2   3   4   5   6.
The composite sample space is given by:

		Sample space for first die					
		1	2	3	4	5	6
Sample	1	2	3	4	5	6	7
space for	2	3	4	5	6	7	8
second	3	4	5	6	7	8	9
die	4	5	6	7	8	9	10
	5	6	7	8	9	10	11
	6	7	8	9	10	11	12

Composite sample
space, i.e. totals
possible when 2 die
are thrown together.

Number of outcomes in composite sample space which are 10 $= N = 3$.
Total number of possible outcomes in composite sample space $= T = 36$.

$$P(\text{Total of 10}) = \frac{N}{T} = \frac{3}{36} = \frac{1}{12}$$

To construct a composite sample space for 3 happenings:
  (*a*) construct a composite sample space for any two of the happenings.
  (*b*) use the results of the composite sample space from (*a*) and the
    sample space for the third happening to construct a composite
    sample space for all 3 happenings.

	Sample space for one of the happenings		Composite sample space for two of the happenings
Sample space for a second happening	Composite sample space for two of the happenings	Sample space for the third happening	Composite sample space for all three happenings

Example: Find the probability that when 2 coins and a die are thrown at the same time the result will be a head, a tail and a number divisible by 3.
The sample space for the die is 1 2 3 4 5 6.
The sample space for one of the coins is Head   Tail.
The composite sample space for the die and one of the coins is given by:

		Die					
		1	2	3	4	5	6
1st Head		H1	H2	H3	H4	H5	H6
Coin Tail		T1	T2	T3	T4	T5	T6

Composite sample space for 1 coin and the die.

Using the composite sample space above and the sample space for the second coin, gives the composite sample space for both coins and the die as follows:

	1st Coin and the Die											
	H1	H2	H3	H4	H5	H6	T1	T2	T3	T4	T5	T6
2nd Head	HH1	HH2	HH3	HH4	HH5	HH6	HT1	HT2	HT3	HT4	HT5	HT6
Coin Tail	TH1	TH2	TH3	TH4	TH5	TH6	TT1	TT2	TT3	TT4	TT5	TT6

Number of outcomes in sample space which have a head, a tail and a number divisible by 3 = N = 4.
Total number of outcomes possible in the sample space = T = 24.

$$P(\text{Head and Tail and number divisible by 3}) = \frac{N}{T} = \frac{4}{24} = \frac{1}{6}.$$

Sample spaces for 4 or more happenings are constructed in a similar manner to that used for 3 happenings.

Tree Diagrams

A small child attempting to draw a tree for the first time usually produces a tree looking remarkably like the one on the right.
The child builds the tree up in stages

by first drawing a trunk, then a cluster of two branches coming out of the trunk, then a cluster of two branches coming out of the ends of each of the first set of branches, and so on.

At each stage a new set of branches is added by drawing a cluster of two branches coming out of the ends of each of the previous set of branches. Tree diagrams, with some modifications, can be used very successfully to find the composite sample space of:

(*a*) combined happenings.
(*b*) repeated happenings.

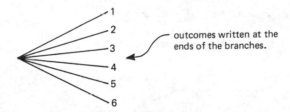

A tree diagram is usually drawn sideways.

The sample space for a *single* happening can be drawn as a cluster of branches, with the possible outcomes written at the ends of the branches.

E.g. the sample space when a coin is tossed is   Head   Tail.

This can be shown as

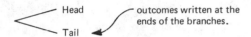

The sample space when a die is tossed is 1   2   3   4   5   6.
This can be shown as

The sample space for a single happening which has:
2 possible outcomes can be shown by a cluster of 2 branches.
6 possible outcomes can be shown by a cluster of 6 branches.
N possible outcomes can be shown by a cluster of N branches.

A tree diagram for a number of happenings or a repeated happening can be built up in stages.

A new set of branches is added to the tree for each new happening or repetition of a happening.

The cluster of branches showing the sample space for the first happening is added to the trunk.

The cluster of branches showing the sample space for the second happening is added to the ends of each of the first set of branches and so on.

For each happening or repetition of a happening the cluster of branches showing the sample space for that happening is added to the ends of the branches of the previous set of branches.

E.g. If 3 coins are tossed one after the other, each coin has a sample space shown by a cluster of 2 branches.

The tree diagram for the first coin is given by adding the cluster

on to the trunk.

For the 2nd coin we have to add the cluster on to the end

of each of the branches of the previous diagram for one coin.

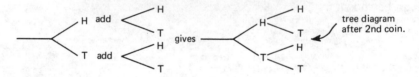

For the 3rd coin we have to add the cluster on to the

ends of each of the branches of the previous diagrams for 2 coins.

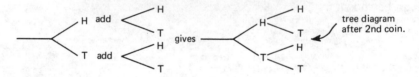

There are various paths leading through a tree diagram starting from the trunk and finishing at the ends of the final set of branches. Consider the tree diagram for 3 coins which is shown on the right in which two such pathways are marked by arrows.

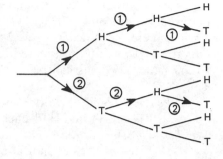

The path marked ① passes through H followed by H followed by T.

The path marked ② passes through T followed by H followed by T.

Path ① gives one of the possible outcomes when three coins are tossed one after the other. This outcome is HHT.

Path ② gives another one of the possible outcomes when three coins are tossed one after the other. This outcome is THT.

Each pathway through a tree diagram gives *one* of the possible outcomes of the combination of happenings represented by the diagram.

A list of all of the possible outcomes is the composite sample space for the combination of happenings.

The composite sample space obtained from a tree diagram can be written on the right of the tree diagram.

The outcome of a particular pathway is written alongside the end of the branch at which it finished.

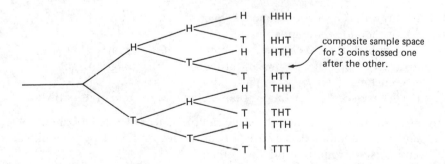

There are occasions when it is necessary to know the order of an outcome. E.g. it may be necessary to know that an outcome occurred in the order Head followed by Head followed by Tail, i.e. HHT.

There are many more occasions when it does *not* matter in what order the outcome occurred.

E.g. if it is only necessary to know how many times 2 heads and 1 tail occurred, the outcomes HHT, HTH and THH are the outcomes of interest. Each of these outcomes involves 2 heads and 1 tail but since order is not important they can be considered to be outcomes of the same type.

Each of the outcomes can be written as H^2T indicating 2 Heads and 1 Tail.

Similarly HHH could be written as H^3 indicating 3 Heads and 0 Tails.

> TTH, THT and HTT each become HT^2 indicating 1 Head and 2 Tails.

> TTT becomes T^3 indicating 0 Heads and 3 Tails.

The tree diagram for 3 coins then becomes

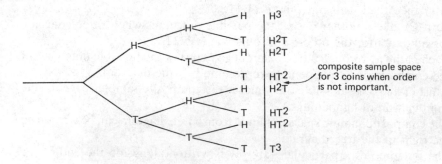

To find the composite sample space for a number of happenings:

    (*a*) draw the clusters of branches which show the sample spaces of each of the *separate* happenings.

    (*b*) combine the clusters from (*a*) into a tree diagram.

    (*c*) trace all of the pathways through the tree diagram, noting the outcome given by each pathway.

    (*d*) list each of the outcomes obtained in (*c*) at the end of its pathway.

Since tree diagrams give the composite sample spaces of when a number of happenings are combined, they are very useful for finding the probabilities of events arising from multiple happenings.

Example: Find the probability that when four coins are tossed together the result is 3 Heads and 1 Tail.

    (*a*) The cluster for all four coins is

(*b–d*) Combining the clusters in a tree diagram gives

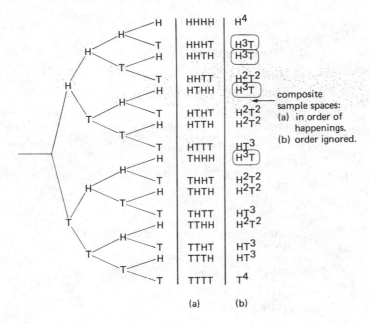

HHHH	H^4	
HHHT	H3T	
HHTH	H3T	
HHTT	H^2T^2	
HTHH	H^3T	composite sample spaces:
HTHT	H^2T^2	(a) in order of happenings.
HTTH	H^2T^2	(b) order ignored.
HTTT	HT^3	
THHH	H^3T	
THHT	H^2T^2	
THTH	H^2T^2	
THTT	HT^3	
TTHH	H^2T^2	
TTHT	HT^3	
TTTH	HT^3	
TTTT	T^4	

(a)   (b)

Number of outcomes in composite sample spaces which are $H^3T = N = 4$.
Total number of possible outcomes in composite sample space $= T = 16$.

$$P(H^3T) = \frac{N}{T} = \frac{4}{16} = \frac{1}{4}.$$

Example: Find the probability of throwing a die and two coins simultaneously, i.e. at the same time, and obtaining an odd number, a head and a tail.

(*a*) The cluster for the die is

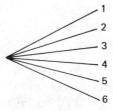

The cluster for both coins is

(*b–d*) Combining the clusters in a tree diagram gives

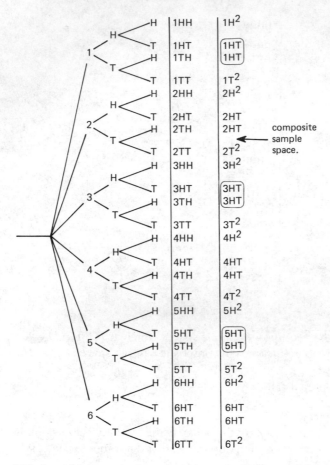

Number of outcomes in composite sample space which give an odd number, a head and a tail = $N = 6$.

Total number of outcomes in composite sample space = $T = 24$.

$$P(\text{Odd} \ \ HT) = \frac{N}{T} = \frac{6}{24} = \frac{1}{4}.$$

Miscellaneous Exercises 6

1. (*a*) A survey on the wearing of seat belts in cars was carried out on Sunday 3rd July 1977, between 2000 hrs and 2010 hrs, on Portstewart Promenade. Of the 200 cars passing the observer, 6 had drivers who were wearing seat belts, and two had front seat passengers who were wearing seat belts.

(i) What is the estimated probability that the driver of the next car would be wearing a seat belt?

(ii) Why could you not make an estimate of the probability of front seat passengers wearing seat belts?

(iii) Give a reason (or reasons) as to why the proportion of drivers wearing seat belts on this occasion was so far below the Road Accident Study Group figure of 15% for drivers and front seat passengers in Northern Ireland.

(*b*) Quite often in fictional stories people decide who should go on a dangerous mission by selecting straws. The person who selects the only long straw is the chosen person. The number of straws is equal to the number of people. Each person selects a straw and holds it hidden until all the straws have been selected, when they are compared.

(i) Use your knowledge of probability to criticise the procedure.

(ii) What alternative method that you have come across in your studies would you suggest to fairly select one person out of six?

NI78

2. Two girls did a survey of traffic at two road junctions by standing at the points *A* and *B* shown.

They found that, of the cars coming from Plum Street,
180 cars turned right into High Road and
270 cars turned left into High Road and that 36 of these then turned right into Hill Street.

(*a*) Using the result of this survey calculate the probability that a car will:

(i) turn left from Plum Street into High Road

(ii) turn left from Plum Street into High Road and then right into Hill Street.

(*b*) Two other girls repeated the survey and found that 230 cars entered High Road from Plum Street and of these 46 turned right. State briefly *two* possible reasons for the different results.

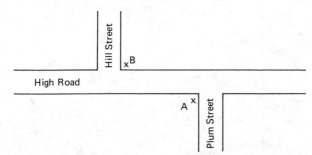

(*c*) Give two *extra facts* which you think the girls should have discovered assuming that they hope to justify the erection of traffic lights at the two junctions.

M77

3. A national survey showed that 75% of the population used P.H. Beans.

   (*a*) If, at random, one person was asked if they used P.H. Beans, what would be the probability that they would say yes?

   (*b*) If, at random, two people were asked if they used P.H. Beans, what would be the probability that they would both say no?   NW78

4. (*a*) (i) Copy and complete the table.

Score	Frequency	F
1	JHT JHT JHT III	
2	JHT JHT JHT	
3	JHT JHT JHT I	
4		19
5		17
6		15

      (ii) If this table shows the results of an experiment involving a die, use them to calculate how often you would expect a six to be thrown in 500 throws of the same die.

     (iii) A group of people decide to obtain an approximation to the probability of a drawing pin landing on its back rather than on its side when thrown.

      Write down what you consider to be the two most important pieces of advice you could give to the group.

  (*b*)

Score	X Class Mid-Value	F Frequency	FX
3–7		2	
8–12		6	
13–17		20	
18–22		46	
23–27		50	
28–32		45	
33–37		22	
38–42		8	
43–47		1	

   (i) Copy and complete the table above.

   (ii) Calculate the mean score.   NI77

Questions 5–8 are concerned with the 'Woof-Bang', a make of car not very well known. 10% of these cars break down before they have covered 100 km. Mr. A and Mr. B each bought one.

5. What is the probability that Mr. A's car will break down before it has been driven 100 km?

6. What is the probability that Mr. B's car will not break down before it has been driven 100 km?

7. What is the probability that neither car will break down before it has been driven 100 km?

8. From the information given, what, if anything, can be said about the probability that Mr. A's car will break down before it has covered 50 km?   EA78

9. What is the probability of choosing a six from a pack of 52 playing cards?   Y78

10. One of the letters of the word SCARBOROUGH is chosen at random. What is the probability that it is:

    (*a*) a vowel?
    (*b*) not S?   Y78

11.

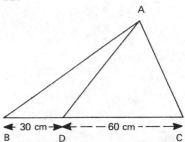

    The diagram represents a triangle *ABC* in which *BD* is 30 cm and *DC* is 60 cm. A person is blind-folded and throws a dart at the triangle. Assuming the dart lands in triangle *ABC*, what is the probability that it also lands in triangle *ABD*?   EA78

12. (*a*) Two players A and B cut a pack of playing cards to decide who should deal the cards for the next game. The player who cuts the higher card will deal, remembering that Aces are the highest cards.

Player A cuts, has a 'ten', and then replaces the card. What is the probability that B will not lose the cut?

(b) What is the probability that a fair coin tossed twice will come down 'heads' both times?   M78

13. Observations at a T-junction showed that of 600 cars, 200 turned left and 400 turned right. Using this information calculate:

   (a) the probability that a car will turn right at the junction,
   (b) the probability that if two cars arrive at the junction both will turn right.   NW77

14. (a) What is the probability of obtaining a 5 with a single throw of a die?
   (b) When two dice are thrown what is the probability of:
      (i) obtaining two 5's?
      (ii) obtaining a score of 10?
      (iii) obtaining a score other than 10?   NW77

15. Two bags each contain six balls identical in size and material, numbered 1 to 6 respectively. The table below shows the total when the number on a ball in one bag is added to the number on any ball in the other bag.

Bag 1 / Bag 2	1	2	3	4	5	6
1	2	3				
2			5	6		
3					8	9
4	5	6				
5			8	9		
6					11	12

Copy the table into your answer book and complete it.
A ball is drawn at random from each bag.

(a) In how many different ways can this be done?
(b) What is the probability that the total score on the two balls is:
   (i) exactly 9,
   (ii) 7 or more,
   (iii) a prime number?   L78

16. Two six-faced die are thrown together. What is the probability of a total score of exactly 11?   Y77

17. (*a*) Find all the *different* results which can be obtained by adding together two *different* digits in the set {1, 2, 3, 4, 5}.

(*b*) If two of the digits are picked at random, what is the probability that their sum will be (i) 7, (ii) 8?

(*c*) What is the probability that the sum of the two numbers picked at random will be odd?   M77

18. (*a*) How many different numbers can be formed by using three different digits from the set {2, 3, 4, 5}?

(*b*) The five digits {1, 2, 3, 4, 5} are placed in a bag, and two of them are picked out unseen. What is the probability that they will both be odd numbers?   M78

19. (*a*) A coin and a die (6 sided) are thrown together.

(i) List all the possible outcomes (e.g., a 3 and a head).

(ii) What is the probability of obtaining

(*A*) an even number with a tail,

(*B*) a number less than 5 with a head?

(*b*) The faces of two cubes *A* and *B* are coloured as follows.

Cube *A*—3 faces are red, 1 face is blue, 1 face is green and 1 face is white.

Cube *B*—1 face is red, 2 faces are blue, 2 faces are green and 1 face is white.

(i) When cube *A* is thrown what is the probability of obtaining a red face?

(ii) When cube *B* is thrown what is the probability of obtaining either a blue face or a green face?   NW75

20. The diagram below, which is incomplete, is designed to show the ways in which the total scores can be combined when two dice, A and B, are thrown together. (For example: the bottom left hand dot represents 1 + 1, the dot above it represents 1 + 2 and the dot beside it represents 2 + 1.)

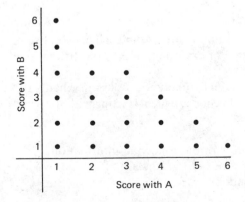

Combinations of total scores when two dice are thrown together (*a*) Copy and complete the 'dot diagram'.

(*b*) Using the diagram, or otherwise, find:

   (i) the total number of ways of combining scores when two dice are thrown together

   (ii) the chance of scoring totals of (1) 5

                                   (2) 10

                                   (3) more than 9

                                   (4) at least 8

   (iii) the chance that, of the two dice thrown, at least one of them is a 6.

(*c*) In two consecutive throws of the two dice together, find the chance of scoring first a total of 4 *followed by* a total of 10.   M78

21. A bag contains 3 blue balls lettered A, B, C respectively. A second bag contains 3 red balls also lettered A, B, C respectively. One ball is picked at random from each bag and all are equally likely to be picked.

(*a*) Copy and complete the following table showing the ways in which the 2 balls may be picked.

Blue	A	A	A	B	B	B		C	C
Red	A	B	C		B		A	B	

(*b*) What is the probability of both blue A and red A being picked?

(*c*) What is the probability of at least one ball lettered B being picked?

(*d*) A third bag contains 3 black balls also lettered A, B, C respectively. One ball is picked at random from each of the three bags. What is the probability of blue A, red A and black A being picked?   WM78

22. The midday menu in a cafe was as follows.

    1st course   soup or fruit juice

    2nd course   curry or salad

    3rd course   trifle or ice cream

(*a*) Assuming that a meal consists of one item from each course, list all the different meals available. For example, soup, curry, trifle is one of the meals available.

(*b*) Assuming that each meal is equally likely to be chosen what is the probability that a customer will choose a meal starting with soup and ending with ice-cream?   M77

Questions 23 and 24 are concerned with the following information.

    There are six cars in a car-park. Four of the cars are black and two are red. Their owners return at random.

23. Find the probability that the first two cars to be driven away will be black. EA78

24. Find the probability that the first two cars to be driven away will be of different colours. EA78

25. (*a*) (i) By writing B for boy and G for girl, write down all the possible ways in which a family of 3 children might occur. (Exclude the possibility of twins or triplets but take notice of the order in which the children are born.)
(ii) How many possible ways are there in which a family of 4 children might occur?
(*b*) In a family of 3 children, what is the probability of
(i) there being 2 boys and 1 girl;
(ii) there being at least one girl. WM76

26. At a school fete organised to raise money for a minibus, the mathematics teacher had made a 'spin the pointer' machine as shown in the diagram below. If the pointer stopped at L, the person lost; if it stopped at W, he won; if it stopped at R, his money was returned.

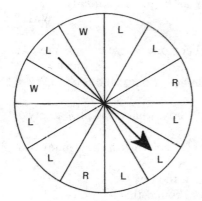

If the twelve sections of the machine were equal in size, calculate:
(*a*) the probability in one spin of
(i) a win,
(ii) a loss,
(iii) a returned coin,
(*b*) the probability in two spins of
(i) two wins,
(ii) two losses,
(iii) a win followed by a loss,

(c) the probability of a win followed by a coin return followed by a loss in three spins.

A boy had ten spins on the machine at 5p per spin. If a win was worth 25p and he had 2 wins and twice his coin was returned, how much profit did he make?   Y78

27. All the pupils in a school were asked to name, from the four listed, their one favourite subject. All did so and their replies are given in the table below.

Chosen subject	Mathematics	Science	French	English
Number	119	85	51	153

(a) How many pupils were there in the school?

(b) If these figures are to be represented on a circular diagram (pie chart) calculate the angle of the sector representing those who chose mathematics.

(c) If one pupil is picked at random what is the probability that he had *not* chosen mathematics? (Assume all pupils are equally likely to have been picked.)   WM77

28. (a) A boy has a box which contains 40 crayons whose colours are red, blue, yellow and green.
  (i) If there is 1 green crayon in the box, what is the probability of selecting a green crayon?
  (ii) If there are 10 red crayons in the box, what is the probability of selecting a red crayon?
  (iii) If the probability of selecting a blue crayon is $\frac{7}{40}$, how many blue crayons are in the box?
  (iv) How many yellow crayons are in the box?

(b) When the boy lost 10 crayons it was found that:
  (i) the probability of selecting a green crayon was 0.
  (ii) the probability of selecting a red crayon was $\frac{3}{10}$.
  (iii) the probability of selecting a blue crayon was $\frac{2}{15}$.
  How many crayons of each colour were lost?   NW78

29. (a) An official is standing at the end of a cross-country race giving out numbered discs to the boys as they finish. Sixty boys have finished and he is left with numbers 61 to 100 in his hand. If he drops them on the ground what is the probability that the first disc he picks up:
  (i) is an even number?

(ii) has a 2 on it?

(iii) has a 6 on it?

(iv) is a perfect square?

(v) is divisible by either 3 or 4?

(*b*) Comment on the truthfulness of the following statement:

One Saturday in Piccadilly, Manchester when both Manchester United and Manchester City were playing at home in cup ties, I asked 500 men what was their favourite sport. 475 said football. This proves that in Manchester, football is the favourite sport for 95% of the male population.

(*c*) I tossed 4 pennies 320 times. If they had formed the perfect distribution how many times would 0 heads, 1 head, 2 heads, 3 heads and 4 heads have occurred?
NW75

7 CORRELATION 1 Scatter Diagrams

When a population is being investigated, data on a number of variables is normally collected, e.g. height, weight, age.

So far we have learned how to process data on one variable at a time, e.g. height.

We can process the data on height and find an average height, and from this a measure of dispersion of the heights.

In a similar way we can process the data on weights and ages.

We can process the data on all the variables, but with our present knowledge we can only consider them separately.

The results obtained from processing the data on each variable may be very useful, but in many investigations it is of much more interest and use to know if there are any relationships between the different variables, e.g.

Are height and weight related in some way?

Is there any connection between smoking and lung cancer?

Are the number of fillings in the teeth of a child affected by the number of sweets that that child eats?

Does regular servicing of a car help to prevent the number of major breakdowns suffered by the car?

To find out whether relationships exist between different variables, the first step is to cross-tabulate the data, i.e. arrange the data in one master table which shows the data for all the variables.

Examples: The following data about 10 girls was collected during a survey.

Name	Mary	Jean	Jane	Ann	Joan	Wendy	Susan	Paula	Alice	June
Age	16	9	11	11	15	14	6	12	15	8

At the same time the following data was collected.

Name	Mary	Jean	Jane	Ann	Joan	Wendy	Susan	Paula	Alice	June
Height	5' 2"	4' 6"	4' 8"	4' 10"	5' 0"	5' 3"	4' 0"	4' 10"	5' 4"	4' 4"

Also at the same time the following data was collected.

Name	Mary	Jean	Jane	Ann	Joan	Wendy	Susan	Paula	Alice	June
Weight (stones)	7	$5\frac{1}{2}$	6	$5\frac{1}{2}$	$6\frac{1}{2}$	$7\frac{1}{2}$	4	$6\frac{1}{2}$	7	$4\frac{1}{2}$

The three separate tables above can be arranged in one master table.

Name	Mary	Jean	Jane	Ann	Joan	Wendy	Susan	Paula	Alice	June
Age	16	9	11	11	15	14	6	12	15	8
Height	5' 2"	4' 6"	4' 8"	4' 10"	5' 0"	5' 3"	4' 0"	4' 10"	5' 4"	4' 4"
Weight	7	$5\frac{1}{2}$	6	$5\frac{1}{2}$	$6\frac{1}{2}$	$7\frac{1}{2}$	4	$6\frac{1}{2}$	7	$4\frac{1}{2}$

From this master table, we can compare the distributions of the three variables age, height and weight.

We can study the table to see if any relationships seem to exist between the variables age, height and weight.

When data has been cross-tabulated, it is often easy to see from a master table that two different variables appear to be related to each other in some way.

It may become apparent that as one variable increases another variable increases, e.g. height and weight.

It may become apparent that as one variable increases another variable decreases, e.g. the age of a girl and the number of outings per month that she makes with her parents.

It may be that some potential relationships between variables are not easy to see from a master table, especially when there are large amounts of data.

A far more satisfactory and safer method of seeking potential relationships is to draw scatter diagrams from the master table.

Drawing scatter diagrams is the 2nd stage in finding out whether relationships exist between different variables.

A *scatter diagram* is a graph of one variable against another variable, e.g. height against weight, age against height, age against weight.

To draw a scatter diagram of two variables:

    (*a*) one of the variables is plotted along the horizontal (across) axis.

    (*b*) the other variable is plotted along the vertical (up) axis.

Example: The heights and weights from our master table of data concerning 10 girls would be plotted as (62", 7), (54", $5\frac{1}{2}$), (56", 6), (58", $5\frac{1}{2}$), (60", $6\frac{1}{2}$), (63", $7\frac{1}{2}$), (48", 4), (58", $6\frac{1}{2}$), (64", 7), (52", $4\frac{1}{2}$).

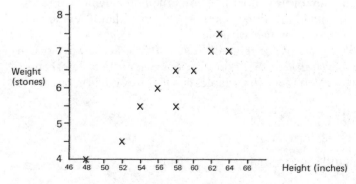

It can be seen from the graph that the plotted points are scattered about the graph, hence the name scatter diagram.

When two variables are plotted against each other, the results will always be a scatter diagram.

On some diagrams the points, although scattered, will form a recognisable pattern with all of the points lying within a narrow band.

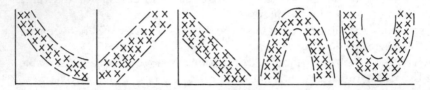

On some diagrams the points will lie on a perfect straight line or a perfect curve.

On some diagrams the points will be so scattered, that by no stretch of the imagination could the points be considered to lie in either a narrow band or on a perfect curve or line.

When a scatter diagram of two variables shows the points lying in a recognisable pattern, the diagram can be taken as evidence that a relationship appears to exist between the two variables.

When two variables appear to be related to each other in some way, we can say that they are *correlated* or that there is a *correlation* between them.

When it has been decided that two variables are correlated it is necessary to decide:

    (*a*) the type of correlation
    (*b*) the degree of correlation

that exists between the two variables.

We will consider only those correlations whose scatter diagrams show the points lying in a narrow *straight* band.

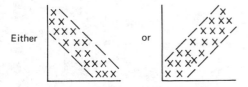

(A) Types of Correlation

(1) Positive or Direct Correlation
If one variable increases in size at the same time as the other variable increases in size, then the correlation between the variables is *positive* or *direct*.

Examples:

Variable 1	Variable 2
age of a man	age of his wife
age of a tree	diameter of the tree
number of cigarettes a person smokes	incidence of lung cancer in that person
height of a child	weight of that child

In each of the above examples, as a general rule, as the size of variable 1 increases then the size of variable 2 also increases.

The scatter diagrams of each of the above examples of positive correlation will have the general shape:

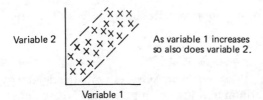

As variable 1 increases so also does variable 2.

(2) Negative or Inverse Correlation
If one variable increases in size at the same time as the other variable

decreases in size, then the correlation between the variables is *negative* or *inverse*.

Examples:

Variable 1	Variable 2
age of a person	liability to catch measles
temperature outside	money spent on heating a house
usage of smokeless fuels	incidence of bronchial disorders
price of oil	number of oil-fired central heating systems fitted

In each of the above examples, as a general rule, as the size of variable 1 increases then the size of variable 2 decreases.

The scatter diagrams of each of the above examples of negative correlation will have the general shape:

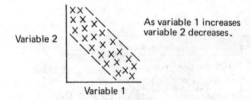

As variable 1 increases variable 2 decreases.

(B) Degree of Correlation

The degree of correlation between two variables is a measure of how highly the two variables are correlated.

The degree of correlation between two variables is called the *correlation coefficient* of the two variables.

The degree of correlation can be approached either descriptively or mathematically.

(*a*) *Descriptively* when correlations are described as
    perfect, high, medium, low or *none*.

Both positive and negative correlations can be broken down into the degrees of correlation: perfect, high, medium and low, giving descriptions such as perfect positive, medium negative, low negative, high positive etc.

(*b*) *Mathematically* when correlations are expressed as a number between −1 and +1.

A negative number indicates a negative correlation.

A positive number indicates a positive correlation.
The closer the number is to either +1 on the one hand and −1 on the other hand, the higher the degree of correlation between the variables.
The following table will act as a *guide*.

Degree of Correlation		Arrangement of points on a scatter diagram
Descriptively	*Mathematically*	
Perfect	−1 or +1	all on a perfect straight line
High	−0.9 or 0.9	all in a very narrow band
Medium	−0.7 or 0.7	all in a narrow band
Low	−0.5 or 0.5	all in a wide to very wide band
None	0	no identifiable pattern whatsoever

Descriptive degrees of correlation are very difficult to judge.
Mathematical degrees of correlation can be worked out in a variety of ways.
We will consider one such way later, i.e. correlation by ranks.

Examples:

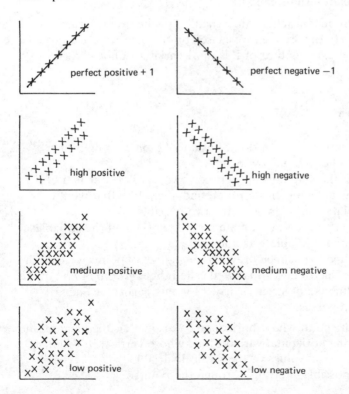

None O

complete absence
of correlation

Absence of correlation would be expected to occur when considering a correlation between the size of a boy's shoes and the boy's mark in an exam, when no correlation would be likely to exist.

Note: A high degree of correlation between two variables does not necessarily mean that the two variables are related to each other.
A high degree of correlation indicates *only* that two variables *may* well be related to each other.
Great care must be taken in the interpretation of degrees of correlation.

Example: The figures for both the number of T.V. sets in use and the number of people dying of heart attacks in the years since 1945 would no doubt show a high positive correlation, but it would be wrong to argue that T.V. sets cause heart attacks. This is an example of false correlation.
As a guide *real* relationships exist if:

    (*a*) changes in one variable cause changes in the other variable.
    (*b*) changes in both variables are caused by changes in a third variable, even if the exact nature of this third variable is not known.

Line of Best Fit

The points on a scatter diagram do not usually lie on a straight line.
The points tend to lie in a narrow band.
When considering the relationship between two variables, it is useful to imagine the points as being clustered around some line that would show graphically the relationship between the two variables.
This line could then be used to estimate the value of one of the variables if the value of the other variable was known.
The best possible line that can be drawn through a set of points is called the *line of best fit*. The line of best fit is often called the *regression line* but the name regression line is not strictly correct for this particular line.
The line of best fit passes:

    (*a*) through the point whose horizontal co-ordinate is the mean of the values of the horizontal variable, and whose vertical co-ordinate is the mean of the values of the vertical variable.
    (*b*) as far as possible centrally through the points.

To draw an approximate line of best fit:

(a) draw a scatter diagram of the two variables.
(b) find the mean of the values of variable 1, i.e. horizontal variable x.
(c) find the mean of the values of variable 2, i.e. vertical variable y.
(d) plot the point whose co-ordinates are given by (b) and (c) and ring this point.
(e) draw a line through the ringed point from (d), in a direction which makes the line pass centrally through the points of the scatter diagram, so that the points are clustered around the line.

Example: The following table shows the marks obtained by ten pupils in French and German tests.

Pupil	A	B	C	D	E	F	G	H	I	J
French	40	56	20	30	48	36	18	24	62	60
German	39	68	16	26	56	40	18	25	74	66

(i) Draw a scatter diagram and a line of best fit for these marks.
(ii) Is there any correlation between the French and German marks? Pupil K obtained 50 in French but missed the German test.
(iii) Use the line of best fit to estimate the mark pupil K could have been reasonably expected to obtain in German, had he been present.

(a) Draw a scatter diagram with French horizontal and German vertical.
(b) For the mean of French marks:

$$(40 + 56 + 20 + 30 + 48 + 36 + 18 + 24 + 62 + 60) \div 10$$
$$394 \div 10 = 39.4$$

(c) For the mean of German marks:

$$(39 + 68 + 16 + 26 + 56 + 40 + 18 + 25 + 74 + 66) \div 10$$
$$428 \div 10 = 42.8$$

(d) Plot the point (39.4, 42.8) and ring it.
(e) Draw the line of best fit through the ringed point from (d).

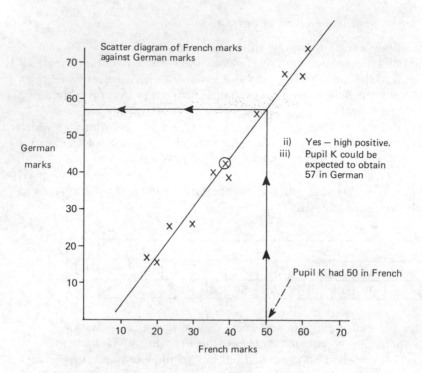

Scatter diagram of French marks against German marks

German marks (y-axis, from 10 to 70)

French marks (x-axis, from 10 to 70)

ii) Yes — high positive.
iii) Pupil K could be expected to obtain 57 in German

Pupil K had 50 in French

Miscellaneous Exercises 7

1. Copy the given axes and sketch a scatter diagram which demonstrates high (but not perfect) positive correlation between two variables.

EA78

2.

$$x = 3 \quad 4 \quad 7 \quad 7 \quad 9$$
$$y = 5 \quad 3 \quad 4 \quad 6 \quad 7$$

The above table shows corresponding values of two variables x and y. If these points were plotted on a scatter diagram, state the co-ordinates of one point through which the line of best fit *must* pass.
(*Note:* You are *not* required to draw the diagram.)   EA78

3. The table gives the examination marks in mathematics and physics for 10 pupils.

Pupil	A	B	C	D	E	F	G	H	I	J
Mathematics mark	73	68	67	40	78	40	88	95	44	74
Physics mark	47	86	47	45	85	30	82	90	14	66

Draw a scatter diagram of these results, on graph paper.
Work out the mean mark in each subject, showing these marks on your scatter diagram.
Draw in a regression line on your graph, and work out the equation of the regression line you have drawn. WM76

4. (*a*) A bookcase contains 100 books, of which 55 are fiction. In random selection what is the probability of selecting two fictional books?
(*b*) Ten pupils are assessed for games ability and graded A, B, C, D or E (A is excellent, E is very poor). Their grades are compared with their mathematics marks.

Games ability grade	A	B	C	C	C	C	C	D	E	E
Mathematics mark	70	76	63	50	43	40	38	25	68	20

(i) Draw a scatter diagram to illustrate the data.
   Use the *x*-axis for games grades (keeping E closest to origin) scale 5 cm between each grade.
   y-axis for mathematics marks 0–100 scale 2 cm representing 10 marks.
(ii) Describe the resulting graph using one of the five statements:
   there is perfect correlation.
   there is very high correlation.
   there is fairly good correlation.
   there is very low correlation.
   there is no correlation.
   If there is correlation state whether it is positive (direct) or negative (inverse). NI78

5. (*a*) (i) Find the standard deviation of the numbers 1, 2, 3, 4, 5 to two significant figures.
   (ii) What does standard deviation measure?
(*b*) An examination consisted of two papers. The marks of ten candidates are shown below.

Candidate	01	02	03	04	05	06	07	08	09	10
Paper I	95	80	67	52	45	40	23	17	15	10
Paper II	87	80	76	70	67	65	61	56	55	53

(i) Draw a scatter diagram on graph paper to illustrate the relationship between the marks.

[Vertical axis: paper I marks, range 0–100, scale 2 cm represents 10 marks. Horizontal axis: paper II marks, range 50–90, scale 4 cm represents 10 marks.]

(ii) Which answer best describes the correlation between the sets of marks? (*a*) High positive, (*b*) high negative, (*c*) low positive, (*d*) low negative, (*e*) zero.   NI77

6. The heights (in cm) and ages (in months) of 15 girls are given in the table.

Pupil	A	B	C	D	E	F	G	H	I	J	K	L	M	N	O
Age (months)	136	140	146	153	161	164	177	179	185	188	189	190	203	206	208
Height (cm)	139	136	151	146	144	162	147	155	163	148	165	162	169	163	175

On graph paper, draw a scatter diagram of these results. Plot age on the *x* axis, and height on the *y* axis, starting both scales at 130.

Label each point plotted with the pupil's letter.

Work out the mean age and the mean height, plotting these results on your graph. Draw in a regression line on your graph.

From your graph, giving reasons for your answers, estimate which girl is the tallest for her age, and which girl is the shortest for her age.   WM77

7. The marks of ten candidates in each of two examinations are given below.

English Language	10	12	20	25	31	34	37	40	44	50
English Literature	12	14	22	22	27	32	31	33	38	37

(i) Plot the points on a scatter diagram, using a scale of 2 cm to represent 10 marks on both axes, and draw in a line of best fit.

(ii) Describe briefly the type of correlation that the diagram appears to suggest exists between the two sets of marks.

(iii) Another candidate scored 27 marks in language but was absent from the literature examination.

Use the diagram to estimate a literature mark for this candidate.   L78

8. The marks obtained by 10 candidates in both history and geography were as follows

Candidate	A	B	C	D	E	F	G	H	I	J
Geography	29	50	80	40	6	58	70	13	61	30
History	37	63	86	50	16	65	81	27	70	41

(*a*) Draw a scatter diagram using the vertical axis for history.

(*b*) On the scatter diagram draw a line of best fit.

(*c*) Estimate

    (i) a Geography mark for a candidate who scored 78 in history.

    (ii) a History mark for a candidate who scored 10 in geography.

(*d*) State the type of correlation shown.   NW78

9.

	Jan.	Feb.	Mar.	April	May	June
Average monthly temperature °C	2.0	3.6	6.4	7.6	11.2	16.8
Number of litres of oil used	550	515	435	380	275	100
	July	Aug.	Sept.	Oct.	Nov.	Dec.
Average monthly temperature °C	17.6	15.6	14.0	13.2	9.6	3.2
Number of litres of oil used	75	150	200	200	310	490

The table shows the number of litres of oil used, for heating, each month by a householder for a year. It also shows the corresponding average monthly temperature.

(*a*) Use the information given to draw a scatter diagram taking 2 cm = 50 litres of oil (vertically) and 2 cm = 2°C (horizontally).

(*b*) On your diagram draw a line of best fit.

(*c*) State the type of correlation shown by the diagram.

(*d*)   (i) Estimate the number of litres of oil used in a month when the average temperature was 8°C.

    (ii) Estimate the average monthly temperature when the householder uses 250 litres of oil.   NW77

10. A reading test was given to ten groups of children, there being 100 children in each group. Each child read out a list of words and a record was kept of the number he read correctly. The average age of the first group was 8 years, that of the second group 8½ years, that of the third group 9 years, and so on. Each group of children was claimed to be a representative cross-section of children of a particular age group. The table below shows the *total* number of words read correctly by the 100 children in each group.

Group	1st	2nd	3rd	4th	5th	6th	7th	8th	9th	10th
Average age, in years	8	$8\frac{1}{2}$	9	$9\frac{1}{2}$	10	$10\frac{1}{2}$	11	$11\frac{1}{2}$	12	$12\frac{1}{2}$
Total number of words correctly read by 100 children	2928	3343	3682	3968	4229	4452	4886	5119	5433	5704

Plot a scatter diagram and draw a line of best fit. Use your line of best fit to estimate the probable number of words correctly read by a child of average reading ability and of age:

(*a*) 8 years 4 months,
(*b*) 10 years 8 months.

Clearly indicate on your diagram any necessary reading made to help you obtain each answer and comment on any assumptions you make.

M76

11. A spring was stretched by the application of weights and the lengths of the spring for various loads were recorded by a student. The results he obtained are given in the table below.

Load (kilograms)	1	2	3	4	5	6	7	8
Length (centimetres)	59.8	62.9	66.5	70.3	73.4	76.7	80.4	84.4

(*a*) Plot a scatter diagram and draw a line of best fit.
(*b*) Use your line of best fit to *estimate* as accurately as possible:
   (i) the length of the spring, in centimetres, when there is no load attached
   (ii) the load, in kilograms, when the length of the spring is 69 centimetres
   (iii) the length of the spring, in centimetres, when the load is 7.4 kilograms.
   Mark your line of best fit at the points where you make readings for these estimates and comment on any assumptions that you make in obtaining each estimate. M77

12. The table below gives the marks scored by ten candidates in the written section and the practical section of a music examination.

Marks in written section	10	14	20	25	30	32	35	38	40	42
Marks in practical section	8	13	18	23	24	25	29	30	32	34

(*a*) Plot a scatter diagram and draw a line of best fit.
Indicating clearly how you obtain each answer
(*b*) use your scatter diagram and line of best fit to estimate:
    (i) the mark that might have been obtained in the practical section
       by a candidate who scored 18 in the written section
    (ii) the mark that might have been obtained in the written section by
       a candidate who scored 20 in the practical section
    (iii) the marks that might have been obtained in each section by a
       candidate whose total score for the two sections was 30.

M78

13. (*a*) The results below show the marks obtained by members of a class of
thirty students in mathematics and physics.

Mathematics	80	40	25	35	55	18	27	55	63	10
Physics	60	30	26	35	60	16	30	40	48	6

Mathematics	73	56	42	15	75	33	28	18	49	12
Physics	60	48	40	15	60	30	10	8	30	15

Mathematics	70	76	56	26	48	42	46	38	35	66
Physics	76	70	46	20	43	20	40	40	33	50

    (i) Draw a scatter diagram for the above information.
    (ii) Draw the best regression line.
    (iii) What score would you expect a student to get in physics if he
       scores 50 in mathematics?
(*b*) In a survey of weekly wages received by a group of workers, the
following results were obtained:

Wage (to the nearest £)	21–30	31–40	41–45	46–50	51–60
Frequency	10	20	20	18	15

Draw a histogram and hence a frequency polygon on the same axes
to show this information. What is the modal class?   EM78

14. The following table gives the monthly totals for exports and imports for
6 months in 1974. These totals are given in million pounds in excess of
£1200 million.

Month	Exports	Imports
1	100	557
2	136	610
3	187	508
4	140	570
5	93	620
6	166	543

(a) Calculate the *actual* monthly average for exports and for imports.

(b) Using a horizontal scale of 2 cm to represent £25 million for exports and a vertical scale of 2 cm to represent £100 million for imports, plot *the figures given in the table* on a scatter diagram.

(c) Draw a suitable straight line to represent the variation of imports with exports.

Comment on the correlation between these items indicated by these figures.   WM76

8 MOVING AVERAGES

Many statistical quantities vary in size as time goes by, e.g. sales figures, unemployment figures, profit figures, hours of sunshine per day.
Quantities which vary in size are called variables.
A set of values that a variable takes over a period of time is called a *time series*.

Day	Mon	Tues	Wed	Thurs	Fri	Sat	Sun	Mon	Tues	Wed
Hours of Sun	4	6	5	8	4	9	8	8	9	7

The set of values, 4 6 5 8 4 9 8 8 9 7, is a time series of the variable hours of sunshine per day. This time series stretches over a ten day period.
A time series graph could be plotted of this data as described on page 53.
The graph of this data is:

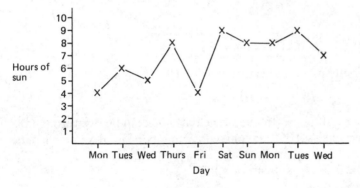

The graph has a jagged appearance, but this graph shows that there was a tendency for the hours of sunshine to increase as time went by.
Most time series graphs have the same sort of jagged appearance, but with an overall trend or tendency for the variable to increase or decrease.

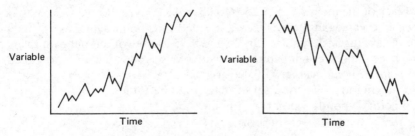

A smooth or at least a smoother graph of time series data helps most people to see the trend of the graph much more clearly.
A time series graph can be smoothed by simply drawing a smooth line centrally through the points on the graph.

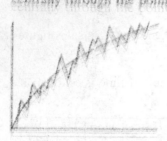

This is adequate for most purposes but other methods exist which are more mathematically precise.
If we take a number of consecutive values from a full set of data and find their mean, we will obtain an average value of a part of the data.
E.g. for the set of data 1   2   5   6   8   9   10   8

Consider 1   2 | 5   6   8   9 | 10   8

The mean of the consecutive values 5   6   8   9, is given by
$$(5 + 6 + 8 + 9) \div 4 = 7.$$

Similarly for 1   2   5   6 | 8   9   10   8 |

The mean of the consecutive values, 8   9   10   8, is given by
$$(8 + 9 + 10 + 8) \div 4 = 8.75$$

If we were to start at the beginning of the data, we could move through the data and average each set of four consecutive numbers, until we reached the end of the data,
giving for 1   2   5   6   a mean of   3.5
          2   5   6   8   ,,   5.25
          5   6   8   9   ,,   7
          6   8   9 10   ,,   8.25
          8   9 10   8   ,,   8.75

Since each of the means, 3.5, 5.25, 7, 8.25 and 8.75, are the averages of four consecutive values at a time, they are called four point averages.
Since the averages, 3.5, 5.25, 7, 8.25 and 8.75, were worked out by moving through the data, averaging four consecutive values at a time, they are called the *4 point moving averages* of the original data.
Other moving averages of a set of data could be found.
E.g. 5 point, 2 point, 3 point or n point.

To find the n point moving averages of a set of data:
starting with the first value in the table of data,

  (a) take n consecutive values in the data.
  (b) add up these n values.
  (c) write the answer to (b) under the middle point of the n values.
  (d) divide the answer to (b) by n.
  (e) write down the answer to (d) under the answer to (c).
  (f) move along the table of values one place, and starting with the
       second value in the table, go back to (a) and repeat the process. The
       process is repeated starting with the third, fourth, and fifth values
       and so on until the mean of the last n values has been found.

The build up of a moving averages table is shown by the example below.

Example: Find the 3 point moving averages of the following time series.
   2   6   4   5   12   10   8   9   16   11
Since a 3 point moving average is required, the values are taken three at a
time.

Stage 1

Values | 2   6   4 | 5   12   10   8   9   16   11

Total of 3 values   Moving Total   12

                                             at the middle of the 3 values

Average of 3 values   Moving Average   4

Stage 2
Values   2 | 6   4   5 | 12   10   8   9   16   11
Moving Total   12   15
Moving Average   4   5

Stage 3
Values   2   6 | 4   5   12 | 10   8   9   16   11
Moving Total   12   15   21
Moving Average   4   5   7

Stage 8: (Final)
Values   2   6   4   5   12   10   8 | 9   16   11 |
Moving Total   12   15   21   27   30   27   33   36
Moving Average   4   5   7   9   10   9   11   12
The working is normally condensed.

Example: Find the five point moving averages of the following time series.

    2 8 10 3 6 11 12 9 15

Since a 5 point moving average is required, the values are taken five at a time.

Values	2	8	10	3	6	11	12	9	15
Moving Totals			29	38	42	41	53		
Moving Average			5.8	7.6	8.4	8.2	10.6		

In the two previous examples of 3 and 5 point moving averages, both 3 and 5 were *odd* and there was a value at the middle point of each part of the data being averaged. The moving totals and averages were placed under that middle value.

When working out *even* pointed moving averages there is no value at the middle point of the part of data being averaged, so the moving totals and moving averages must be placed *between* the two middle values, at each stage.

Example: Find the 4 point moving averages of the following time series.

1 2 5 6 8 9 10 8

Value	1	2	5	6	8	9	10	8
Moving Total		14	21	28	33	35		
Moving Average		$3\frac{1}{2}$	$5\frac{1}{4}$	7	$8\frac{1}{4}$	$8\frac{3}{4}$		

Exercise 8a:

1. Find the 2 point moving averages of the following series.

   2 4 6 4 8 8 6 4 8 10 6 4

2. Find the 3 point moving averages of the following series.

   10 11 6 9 10 5 8 9 4

3. Find the 4 point moving averages of the following series.

   8 11 14 10 5 7 10 7 3 6 8 5

4. Find the 5 point moving averages of the following series.

40 44 46 28 33 34 38 23 24 15 28 21

5. Find the 6 point moving averages of the following series.

20 22 24 24 29 32 26 28 30 21 25 27 27 32 36

Another way of smoothing a time series graph is to plot a graph of the moving averages of the series.

Each moving average must be plotted against the middle of the time interval which it covers.

The points on the graph are connected with *straight lines*.

Example: Find the 2 point moving averages of the time series shown below and plot them on a graph.

Week	1	2	3	4	5	6	7	8	9	10
Rainfall cm	4	6	5	8	5	9	8	6	10	9

For the moving averages

Week	1	2	3	4	5	6	7	8	9	10
Cm of rain	4	6	5	8	5	9	8	6	10	9
Moving Totals		10	11	13	13	14	17	14	16	19
Moving Averages		5	$5\frac{1}{2}$	$6\frac{1}{2}$	$6\frac{1}{2}$	7	$8\frac{1}{2}$	7	8	$9\frac{1}{2}$

Note: The moving averages must be plotted against the middle of the time intervals, i.e. at $1\frac{1}{2}$, $2\frac{1}{2}$, $3\frac{1}{2}$ weeks and so on.

In the graph below, the time series is shown by a solid line, the moving averages are shown by the dotted line.

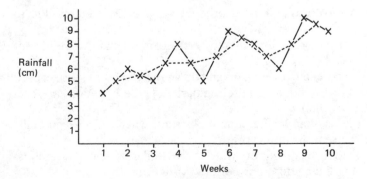

194   *Basic Statistics*

Exercise 8b:

1. Plot a graph of the series given in question 1 of Exercise 8a. On the same graph plot the 2 point moving averages of the series.

2. Plot a graph of the series given in question 2 of Exercise 8a. On the same graph plot the 3 point moving averages of the series.

3. Plot a graph of the series given in question 3 of Exercise 8a. On the same graph of the series plot the 4 point moving averages of the series.

4. Plot a graph of the series given in question 4 of Exercise 8a. On the same graph plot the 5 point moving averages of the series.

5. Plot a graph of the series given in question 5 of Exercise 8a. On the same graph plot the 6 point moving averages of the series.

Graphs showing moving averages are very useful in attempting to predict future events.
E.g. a firm with a sales graph showing an upward trend might decide to expand production to meet a predicted increase in sales.
Another firm with a profits graph showing a downwards trend might decide to close down before a predicted bankruptcy.
The shape of a time series graph depends on the factors which affect a time series.
A time series is affected by:

    (*a*) overall trends, i.e. secular (long term) variations.
    (*b*) periodic variations, i.e. seasonal and cyclical variations.
    (*c*) random variations.

An *overall trend* is shown by the tendency for a time series to rise or fall steadily.
E.g. the costs involved in making a product tend to rise steadily due to inflation.
The profits of a firm tend to drop steadily if the firm is run in an inefficient manner.
Periodic variations, i.e. movements up and down in a time series, are caused by factors which recur with a definite regularity. These variations are predictable because of their periodicity
E.g. the sales figures of an ice cream manufacturer will show rises and falls depending on the seasons of the year.
The costs involved in heating a building show similar seasonal variations.
Tidal heights show many different periodic variations.

The use of *n*-point moving averages enables periodic variations in a time series to be smoothed out, in order to enhance the overall trend of the series. The value of *n* chosen to smooth out periodic variations should be a multiple of the period of the variations.

E.g. if similar variations in a time series recur at each fourth value in the series, then either 4 point, or 8 point, or 12 point etc. moving averages of the series should be computed.

The larger the value of *n* chosen, the greater the smoothing effect the moving averages will have on the series.

Random variations in a time series are caused by factors which occur at random. These variations are not predictable.

E.g. strikes and machinery breakdowns cause variations in the production figures of a firm.

Storm surges cause unpredictable variations in tidal heights.

Example: The profits, in thousands of pounds, of the XYZ Bedding Company over a period of 15 years are shown in the table below.

Year	1964	1965	1966	1967	1968	1969	1970	1971
Profit	42	53	42	49	64	72	83	67

Year	1972	1973	1974	1975	1976	1977	1978
Profit	74	94	82	103	92	99	114

Find the 5 yearly moving averages of the profits and plot them on a graph. The table below shows the calculations involved in calculating the 5 yearly moving averages.

Year	1964	1965	1966	1967	1968	1969	1970	1971
Profit	42	53	42	49	64	72	83	67
Mov. Totals			250	280	310	335	360	390
Mov. Averages			50	56	62	67	72	78

Year	1972	1973	1974	1975	1976	1977	1978
Profit	74	94	82	103	92	99	114
Mov. Totals	400	420	445	470	490		
Mov. Averages	80	84	89	94	98		

The graph showing the time series of the profits (solid line) and the moving averages of the profits (dotted line) is shown below.

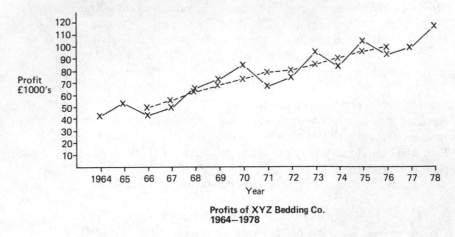

Profits of XYZ Bedding Co.
1964–1978

Example:
The cost in pounds of the electricity bills of a family for each quarter during the years 1978, 1979 and 1980 are shown in the table below.

		Quarter			
		1st	*2nd*	*3rd*	*4th*
	1978	36	18	19	25
Year	1979	30	19	20	24
	1980	38	18	21	23

Find the four quarterly moving averages of the bills and plot them on a graph.
The table below shows the calculations involved in calculating the four quarterly moving averages.

Year Quarter	1978				1979				1980			
	1st	*2nd*	*3rd*	*4th*	*1st*	*2nd*	*3rd*	*4th*	*1st*	*2nd*	*3rd*	*4th*
Cost £'s	36	18	19	25	30	19	20	24	38	18	21	23
Mov. Total			98	92	93	94	93	101	100	101	100	
Mov. Average			$24\frac{1}{2}$	23	$23\frac{1}{4}$	$23\frac{1}{2}$	$23\frac{1}{4}$	$25\frac{1}{4}$	25	$25\frac{1}{4}$	25	

The graph showing the times series of the costs of the bills (solid line) and the moving averages of the bills (dotted line) is shown below.

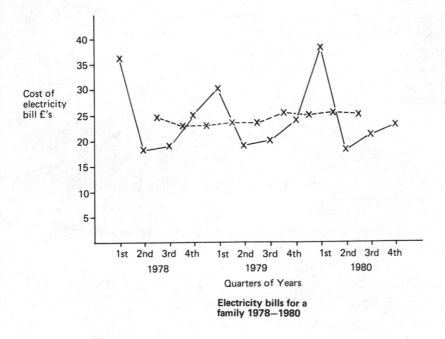

Electricity bills for a family 1978–1980

Exercise 8c:

1. Choose a suitable n point moving average for each of the time series given in Exercises 3f.

2. Calculate the chosen n point moving averages for each of the time series given in Exercises 3f.

3. Plot each of the time series given in Exercises 3f and on the same graph plot the n point moving averages for the series.

Miscellaneous Exercises 8

1. What is a time series? WM75

2. Calculate the 5 point moving averages for   10   11   9   13   8   7   4
WM75

3. Calculate the 5 point moving averages for 12   16   9   11   17   13   15
WM76

Questions **4** and **5** concern the following statistics which show the number of houses sold by an agent during the first six months of 1977.

Jan	Feb.	Mar.	April	May	June
14	9	13	17	21	23

4. What is the value of the second 3-monthly moving average? EASY

5. If the fifth 3-monthly moving average is 21, how many houses were sold during July? EASY

6. The second four point moving average of 11 16 9 x 12 15 is 11. Find the value of x. WM77

7. The following figures show the annual sales of a certain product over a 15 year period.

Year	1	2	3	4	5	6	7	8	9	10	11	12	13	14	15
Sales	350	360	370	420	480	500	490	510	520	550	550	540	520	500	590

(i) Draw a graph showing the annual sales over this 15 year period.
(ii) Work out the 5 yearly moving averages over this period.
(iii) Using the same axes as for the first graph, draw a graph of the 5 yearly moving average. WM75

8. The following figures represent the monthly output of a certain manufacturing company in each month of a 12 months period.

Month	1	2	3	4	5	6	7	8	9	10	11	12
Output	4.5	6.2	7.6	5.6	5.7	6.8	8.7	6.6	5.8	8.3	9.1	7.3

(i) Draw a graph showing the monthly output over this 12 month period.
(ii) Work out the 3 monthly moving averages over this period.
(iii) On the same axes as the first graph, draw a graph of the 3 monthly moving average.
(iv) Estimate the output for the next month, i.e. month 13. WM76

9. The following figures are quarterly unemployment percentages in the building trade.

Quarter	1	2	3	4	5	6	7	8	9	10	11	12
%Unemployed	15.5	11.8	12.1	13.7	14.6	11.1	11.1	12.3	14.4	9.8	9.7	11.5

(a) Draw a graph showing the variation in unemployment percentages over the period of 12 quarters.
(b) Work out the four-quarterly moving averages over this period.
(c) On the same axes as the first graph, draw a graph of the four-quarterly moving averages.
(d) Estimate the percentage unemployed for the next quarter after the period covered by the above data, showing your working. WM77

10. The figures below show the amounts (in millions of pounds) spent on food in a certain country during twelve consecutive quarters.

	First Year	Second Year	Third Year
First quarter	745	821	911
Second quarter	799	872	973
Third quarter	845	907	989
Fourth quarter	864	940	999

(a) By using as large a scale as possible and by marking the lowest point on the 'amount spent' axis as 700 millions of pounds, draw a graph to illustrate the data in the above table.
(b) Calculate and tabulate the four-quarterly moving averages.
(c) *Superimpose* on your first graph a graph of the four-quarterly moving averages.
(d) Comment critically on these graphs. M76

11. The following table shows the number of children in a school involved in some form of special school journey during each of the three terms for the last four years.

	1st term	2nd term	3rd term
1st year	207	345	180
2nd year	249	366	186
3rd year	261	396	210
4th year	291	426	243

(a) By using as large a scale as possible and by making the lowest point on the 'number of children' axis 200, draw a graph to illustrate the data in the table above.
(b) Calculate and tabulate the three-termly moving averages.
(c) *Superimpose* on your first graph a graph of the three-termly moving averages.

(d) Describe the trend indicated by the moving averages graph, giving possible reasons for such a trend. M77

12. The following table shows the number of pairs of shoes sold in a certain shop on each day of the week (Monday to Saturday) over a period of three weeks.

	Week 1	Week 2	Week 3
Monday	10	22	22
Tuesday	30	42	42
Wednesday	10	22	34
Thursday	30	42	42
Friday	40	52	54
Saturday	60	72	96

(a) Draw a graph to illustrate the data in the table above.
(b) Calculate and tabulate the six-day moving averages.
(c) *Superimpose* on your first graph a graph of the six-day moving averages.
(d) Comment on the trend indicated by the moving averages graph. M78

13. The table below gives the number of passengers (in thousands) carried by a certain cross-channel ferry during the first, second, third and fourth quarter of each year for the three years 1975, 1976 and 1977:

Year	Quarters			
	1	2	3	4
1975	2	4	7	3
1976	3	5	8	2
1977	3	6	10	4

(i) Plot the line graph of the above time-series.
(ii) Calculate the four-quarterly moving averages.
(iii) Plot the four-quarterly moving averages on the same diagram as in part (i).
(iv) Comment on any inference that can be drawn from these graphs.
(v) State how a suitable moving average is chosen. EM78

14.

Year	1st quarter	2nd quarter	3rd quarter	4th quarter
1975	33.8	26.9	28.4	36.9
1976	38.9	31.5	33.0	42.3
1977	45.2	37.0	38.7	48.8

The table above shows the electrical power used (in millions of units), for street lighting in a certain town, in each quarter of three successive years.

(a) Draw a graph showing the variation in electrical power used over the period of twelve quarters.

(b) Work out the four-quarterly moving averages over this period.

(c) On the same axes as the first graph, draw a graph of the four-quarterly moving averages.

(d) Use the graph obtained in (c) to estimate the electrical power used in the first quarter of 1978. Be sure to show your working.   WM78

15. Road Casualties in Great Britain (thousands)

Year	1974	1975	1976	1977
1st quarter	68	71	72	76
2nd quarter	78	78	82	
3rd quarter	87	86	80	
4th quarter	89	89	98	

Source: Department of the Environment
Scottish Development Department.

Draw a graph to illustrate the above figures. Start your vertical axis at 60 thousand casualties and use a scale of 5 cm to represent 10 thousand casualties. Space the quarters at 1 cm intervals on the horizontal axis.
Calculate the four-quarterly moving averages and superimpose the graph of these moving averages on the first graph.
With the help of the second graph, make a calculated estimate of the likely number of road casualties in the second quarter of 1977. How reliable is such an estimate likely to be?   EA78

16. Give an example of statistics forming a time series which are likely to show a seasonal variation.   EA78

17. State one important reason for using moving averages.   WM75

18. Explain what is meant by secular variation in a time series   WM76

9 CORRELATION 2 Ranks

Correlation by Ranks

If the values of a variable are considered in order of size, then each value can be given a *rank* i.e. a place in an order of merit.
The greatest values are given the highest rankings 1, 2, 3 . . . . . . , the smallest values are given the lowest rankings.
E.g. the French marks of the example on page 181.

Pupil	A	B	C	D	E	F	G	H	I	J
Mark	40	56	20	30	48	36	18	24	62	60

Pupil I who was first with the highest mark of 62 is given the rank of 1.
Pupil J who was second with a mark of 60 is given the rank of 2.
Pupil G who was tenth with a mark of 18 is given the rank of 10.
The French rankings are as follows:

Pupil	A	B	C	D	E	F	G	H	I	J
French Rank	5	3	9	7	4	6	10	8	1	2

The German marks of the example on page 181.

Pupil	A	B	C	D	E	F	G	H	I	J
Mark	39	68	16	26	56	40	18	25	74	66

Pupil I with a mark of 74 is given the rank of 1.
Pupil B with a mark of 68 is given the rank of 2. etc.
The German rankings are as follows:

Pupil	A	B	C	D	E	F	G	H	I	J
German Rank	6	2	10	7	4	5	9	8	1	3

The two sets of rankings arranged in the same table give:

Pupil	A	B	C	D	E	F	G	H	I	J
French Rank	5	3	9	7	4	6	10	8	1	2
German Rank	6	2	10	7	4	5	9	8	1	3

It is apparent from this table that some pupils do not have the same ranks in both subjects i.e. there is a *difference* in their rankings in the two subjects. E.g. Pupil B has a rank of 3 in French and a rank of 2 in German. There is a difference in pupil B's rankings of $3 - 2 = 1$. Pupil A has a rank of 5 in French and a rank of 6 in German. There is a difference in pupil A's rankings of $5 - 6 = -1$. For each pupil the difference in his rankings can be obtained by subtracting his German rank from his French rank. This gives:

Pupil	A	B	C	D	E	F	G	H	I	J
French Rank	5	3	9	7	4	6	10	8	1	2
German Rank	6	2	10	7	4	5	9	8	1	3
Difference in Ranks	−1	1	−1	0	0	1	1	0	0	−1

The differences in the ranks of two variables can be used to find the *degree of correlation* between the variables.
One method of finding the degree of correlation, or the *correlation coefficient*, between 2 variables was developed by C. Spearman in 1906.
If each item in a set of data has a pair of ranks then *Spearman's coefficient of correlation by ranks*, R, is defined by:

$$\text{either } R = 1 - \frac{6 \, \Sigma \, D^2}{n(n^2-1)}$$

where D is the difference in ranks of an item,
and n is the total number of items, i.e. pairs of ranks, in the set of data.

$$\text{or } R = 1 - \frac{6 \times \text{the total of the squares of the differences in ranks}}{\text{number of items} \times (\text{the square of the number of items} - 1)}$$

For the French and German marks we have:

The differences in ranks D are: $-1 \quad 1 \quad -1 \quad 0 \quad 0 \quad 1 \quad 1 \quad 0 \quad 0 \quad -1$.

The squares of the differences in ranks D^2 are:

$$(-1)^2 \quad 1^2 \quad (-1)^2 \quad 0^2 \quad 0^2 \quad 1^2 \quad 1^2 \quad 0^2 \quad 0^2 \quad (-1)^2$$

This gives: 1 1 1 0 0 1 1 0 0 1
The total of the squares of the differences in ranks ΣD^2 is
$$1+1+1+0+0+1+1+0+0+1=6.$$
The number of items (n), in this case pupils, is 10.

$$R = 1 - \frac{6 \times 6}{10(10^2-1)}$$

$$= 1 - \frac{36}{10(100-1)} = 1 - \frac{36}{10(99)} = 1 - \frac{36}{990}$$

$$= 1 - 0.03\overset{..}{6}3\overset{..}{6}$$
$$= 0.96 \text{ to 2 decimal places.}$$

To find the coefficient of correlation by ranks:

(a) make a table with seven columns and head them as shown below.

Item	Values of Variable 1	Values of Variable 2	Ranks of Variable 1	Ranks of Variable 2	D Diffs. in Ranks	D² (Diffs)²

(b) fill in the first 3 columns.
(c) rank the values of variable 1 and put the ranks in the fourth column, i.e. ranks of variable 1.
(d) rank the values of variable 2 and put the ranks in the fifth column, i.e. ranks of variable 2.
(e) for each line of the table subtract the rank of variable 2 from the rank of variable 1.
(f) put the answer to (e) in the sixth column, i.e. difference in ranks or D.
(g) for each line of the table square the number in the sixth column, i.e. difference in ranks or D.
(h) Put the answers to (g) in the seventh column, i.e. (differences in ranks)² or D².
(i) add up the seventh column. This gives the total of the squares of the differences in ranks ΣD^2.
(j) find the number of items (n), that is, the number of pairs of ranks.
(k) use the answers to (i) and (j) in the following formula:

$$R = 1 - \frac{6 \Sigma D^2}{n(n^2-1)}$$

(l) write down the answer to (k) as the coefficient of correlation by ranks.

Example: The marks in mathematics and cookery for a group of 10 children are shown below. Find the coefficient of rank correlation between the mathematics and cookery marks.

Pupil	A	B	C	D	E	F	G	H	I	J
Maths.	90	15	29	83	81	41	44	73	53	68
Cookery	74	37	65	45	23	50	32	55	81	72

Rewriting the above table we obtain:

Pupil	Maths. Marks	Cookery Marks	Maths. Rank	Cookery Rank	D Difference Ranks	D² (Difference Ranks)²
A	90	74	1	2	−1	1
B	15	37	10	8	2	4
C	29	65	9	4	5	25
D	83	45	2	7	−5	25
E	81	23	3	10	−7	49
F	41	50	8	6	2	4
G	44	32	7	9	−2	4
H	73	55	4	5	−1	1
I	53	81	6	1	5	25
J	68	72	5	3	2	4

$$\Sigma D^2 = 142$$

Total of squares of differences

Number of pupils $n = 10$.

$$R = 1 - \frac{6 \, \Sigma \, D^2}{n(n^2-1)}$$

$$= 1 - \frac{6 \times 142}{10(10^2-1)} = 1 - \frac{852}{10(10^2-1)}$$

$$= 1 - \frac{852}{10(99)} = 1 - \frac{852}{990}$$

$$= 1 - 0.860$$

$$= 0.14$$

A variable may have more than one item with the same value, i.e. there may be tied items.

e.g.

Pupil	A	B	C	D	E	F	G	H	I	J
Music Marks	72	65	60	42	60	55	42	50	35	42

Pupils C and E have the same mark of 60.
Pupils D, G and J have the same mark of 42.
In a situation such as the one above, it is usual (though not mathematically correct) to give tied items the same rank as each other.
The rank given to each of a set of tied items is the arithmetic mean of the ranks which the items would have been given if their values had been slightly different, i.e. C with 61 and E with 60 perhaps.
C and E would have had the ranks of 3 and 4.

Both C and E are given the rank $\dfrac{3+4}{2} = \dfrac{7}{2} = 3\frac{1}{2}$.

D, G and J would have had the marks of 7, 8 and 9.

Each of D, G and J is given the rank $\dfrac{7+8+9}{3} = \dfrac{24}{3} = 8$.

Example: Find the coefficient of rank correlation between the music and history marks shown in the table below.

Pupil	A	B	C	D	E	F	G	H	I	J
Music Mark	72	65	60	42	60	55	42	50	35	42
History Mark	55	42	50	62	70	50	40	55	30	35

Rewriting the above table we obtain:

Pupil	Music Mark	History Mark	Music Rank	History Rank	D	D²
A	72	55	1	$3\frac{1}{2}$	$-2\frac{1}{2}$	$6\frac{1}{4}$
B	65	42	2	7	-5	25
C	60	50	$3\frac{1}{2}$	$5\frac{1}{2}$	-2	4
D	42	62	8	2	6	36
E	60	70	$3\frac{1}{2}$	1	$2\frac{1}{2}$	$6\frac{1}{4}$
F	55	50	5	$5\frac{1}{2}$	$-\frac{1}{2}$	$\frac{1}{4}$
G	42	40	8	8	0	0
H	50	55	6	$3\frac{1}{2}$	$2\frac{1}{2}$	$6\frac{1}{4}$
I	35	30	10	10	0	0
J	42	35	8	9	-1	1

$$\Sigma D^2 = 85$$

Number of pupils, $n = 10$.

$$R = 1 - \frac{6\,\Sigma\,D^2}{n(n^2-1)} = 1 - \frac{6 \times 85}{10(10^2-1)} = 1 - \frac{510}{10(100-1)}$$

$$= 1 - \frac{510}{990} = 1 - 0.515\dot{1}$$

$$= 0.48 \text{ to 2 decimal places}$$

The coefficient of rank correlation between two variables can be calculated even if the values of the variables are not known.

All that need be known are the *ranks* of the variables.

Qualities such as colour, enjoyment, happiness and so on, are difficult or even impossible to measure accurately.

If a quality can be ranked, e.g. colour by degree of darkness, then coefficients of rank correlation involving qualities can be calculated.

E.g. a rank correlation between the amount of money spent in making films and the enjoyment felt by people seeing the films could be calculated.

Example: A film critic compared his opinions of five films with the cost of making the films. The results are shown in the table below.

Film	W	W	X	Y	Z
Cost (millions of £'s)	12	7	1	4	2.5
Critic's Opinion	Very Good	Poor	Good	Excellent	Good

To find the coefficient of correlation by ranks we have:

Film	Cost	Critic's Opinion	Cost Ranking	Critic Ranking	D	D²
W	12	Very Good	1	2	−1	1
W	7	Poor	2	5	−3	9
X	1	Good	5	3	2	4
Y	4	Excellent	3	1	2	4
Z	2.5	Good	4	3	1	1

$$\Sigma D^2 = 16$$

number of films, $n = 5$

$$R = 1 - \frac{6 \Sigma D^2}{n(n^2 - 1)} = 1 - \frac{6 \times 16}{5(5^2 - 1)} = 1 - \frac{99}{5(24)}$$

$$= 1 - \frac{99}{120} = 1 - 0.825$$

$$= 0.175$$

Miscellaneous Exercise 9

1. The following examination marks are to be placed in rank order so that a rank correlation coefficient may be calculated. In the table below insert the ranks corresponding to the marks.

Examination Marks	57	42	65	25	42	51	65	39	65	42
Rank										

WM78

2. In an examination a student was asked to calculate the correlation between the number of television sets sold and the number of cinema attendances. He calculated the correlation coefficient as −2.1. Explain why his answer could not have been correct. EA78

3. Five candidates took examinations in English and mathematics, obtaining marks as follows:

Candidate	A	B	C	D	E
English Marks	32	43	18	21	45
Maths Marks	31	30	43	38	18

Write down the rank correlation coefficient between the two sets of marks.   EA78

4. The following table gives the examination marks in English and mathematics of 10 pupils.

Pupil	A	B	C	D	E	F	G	H	I	J
English mark	24	27	35	37	38	44	51	53	62	73
Mathematics mark	37	24	22	74	18	33	61	41	48	64

(i) Calculate the coefficient of rank correlation between the two subjects.
(ii) Plot the marks for the two subjects on a scatter graph. Draw in the line of best fit. Use this line to estimate the mathematics mark corresponding to an English mark of 47.   WM75

5. The following table gives the examination marks for 10 pupils in mathematics, paper 1 and paper 2.

Pupil	A	B	C	D	E	F	G	H	I	J
Paper 1	16	27	34	53	49	36	60	64	72	75
Paper 2	23	22	44	57	75	65	57	72	90	61

(a) Calculate the coefficient of rank correlation between the two papers.
(b) Draw a scatter diagram of the marks, on graph paper. (Use the *x* axis for paper 1 marks, and the *y* axis for paper 2 marks). Label each point plotted with the pupil's letter.
(c) Work out the mean mark for each paper and plot your results.
(d) Draw in a line of best fit.
(e) Use this line to estimate a mark for paper 2 for a pupil who scored 60 marks on paper 1 and was absent for paper 2.   WM78

6. The weights, correct to the nearest kilogram, and the heights, correct to the nearest centimetre, of 12 adults are given in the table below.

Adult	A	B	C	D	E	F	G	H	I	J	K	L
Weight in kg	74	70	55	58	51	56	57	62	62	69	62	67
Height in cm	175	173	164	171	152	164	155	169	168	178	186	175

(a) Work out the *rank* orders of the weights and of the heights of these 12 adults, give rank 1 to the heaviest weight and the tallest height. *Indicate clearly* how you obtained the ranks when the weights or heights of two or more adults are equal.

(b) Calculate the coefficient of rank correlation between the weights and heights of these adults.

(c) State, with reasons, what significance, if any, you attach to your result.   M76

7. The table below show the best performance of 16 athletes in the 100 metres and the 400 metres.

Athlete	100 metres (seconds)	400 metres (seconds)
A	10.7	46.7
B	10.5	45.6
C	10.8	46.9
D	10.5	46.6
E	10.5	46.2
F	10.9	48.0
G	10.2	46.3
H	10.6	46.8
I	10.4	46.5
J	10.8	48.4
K	10.8	47.2
L	10.3	46.4
M	10.8	47.6
N	10.7	45.9
O	10.4	46.4
P	10.6	47.5

(*a*) Work out the rank orders of the times for the 100 metres and the 400 metres, in each case giving rank 1 to the *fastest* time. When working out the rank orders, explain *clearly* how you make the calculations when two or more athletes have the same time.

(*b*) Calculate the coefficient of rank correlation between the ranks for the 400 metres times.

(*c*) What significance, if any, do you attach to the value of the coefficient you have obtained? Add any critical comments you wish.   M77

8. The points awarded to 12 competitors in a school gymnastics competition by two judges, X and Y, are given below.

Competitor	Judge X	Judge Y
A	5.0	5.1
B	5.5	5.5
C	5.4	5.3
D	5.6	5.4
E	5.2	5.2
F	5.8	5.7
G	5.7	5.6
H	5.9	6.0
I	5.4	5.2
J	5.6	5.5
K	5.1	5.0
L	5.6	5.5

(*a*) Work out the *rank* order of the points awarded by each judge, giving rank 1 to the highest number of points awarded. *Indicate clearly* how you obtain ranks when the points awarded to two or more competitors are equal.

(*b*) Calculate the coefficient of rank correlation between the points awarded by the two judges.

(*c*) What significance, if any, do you attach to the value of the coefficient in this case?   M78

9.

British Rail – Passenger Receipts and Traffic

Year	Fares paid by Passengers (£ millions)	No. of Passenger Journeys (millions)
1966	179	835
1967	180	837
1968	185	831
1969	205	805
1970	228	824
1971	261	816
1972	274	754
1973	297	728
1974	329	732
1975	429	715

Source: Dept. of the Environment.

Calculate a rank correlation coefficient between passenger fares paid and the number of passenger journeys made.

State briefly the meaning of your result and give two possible reasons for it. EA78

10 USE OF GUESSED MEAN

Mean and Standard Deviation: Further Study

If a number is subtracted from another number, the result is called the *difference* between the two numbers.
Consider the array A below.
Array A is 101, 100, 108, 102, 104.
An array B can be formed by subtracting 100 from every value in Array A.
Array B is 1, 0, 8, 2, 4.
Note: Array B consists of the *differences* obtained when the number 100 is subtracted from each of the values in array A.
Finding the mean of both arrays by the method on page 94 gives:

For Array A:
(*a*) there are 5 values.
(*b*) $101+100+108+102+104=515$.
(*c*) $515 \div 5=103$.
(*d*) mean$=103$.

For Array B:
(*a*) there are 5 values.
(*b*) $1+0+8+2+4=15$.
(*c*) $15 \div 5=3$.
(*d*) mean$=3$.
but $103=3+100$

Notice it appears that:
The Mean of Array A = The Mean of Array B + 100 ← this number was subtracted from all of array A to form array B.
But Array B is an array of *differences*.
So the Mean of Array A = The Mean of the array of differences +100.

Other arrays of differences can be found from array A.
Array C   3,   2, 10,   4,   6   by subtracting   98 throughout array A.
Array D   51, 50, 58, 52,   54   by subtracting   50 throughout array A.
Array E -4, -5,   3, -3, -1 by subtracting 105 throughout array A.
Array F -1, -2,   6,   0,   2   by subtracting 102 throughout array A.
Array G -9, -10, -2, -8, -6 by subtracting 110 throughout array A.

For Array C:
(*a*) 5 values.
(*b*) $3+2+10+4+6=25$.
(*c*) $25 \div 5=5$.
(*d*) mean$=5$.
but $103=5+98$.

For Array D:
(*a*) 5 values.
(*b*) $51+50+58+52+54=265$.
(*c*) $265 \div 5=53$.
(*d*) mean$=53$.
but $103=53+50$.

For Array E:
(*a*) 5 values.

For Array F:
(*a*) 5 values.

(b) $-4+(-5)+3+(-3)+(-1)=-10$. (b) $-1+(-2)+6+0+2=5$.
(c) $-10 \div 5 = -2$. (c) $5 \div 5 = 1$
(d) mean $= -2$. (d) mean $= 1$.
but $103 = -2 + 105$. but $103 = 1 + 102$.

For Array G:
(a) 5 values.
(b) $-9+(-10)+(-2)+(-8)+(-6)=-35$.
(c) $-35 \div 5 = -7$.
(d) mean $= -7$.
but $103 = -7 + 110$.

Notice the mean of array A = mean of array C + 98 ⎫ These are the numbers
        or = mean of array D + 50 ⎪ which were subtracted
        or = mean of array E + 105 ⎬ from all of array A
        or = mean of array F + 102 ⎪ in order to form
        or = mean of array G + 110 ⎭ arrays C, D, E, F, G.

Note: Arrays B, C, D, E, F and G are all arrays of differences, but no matter what number was subtracted from all of array A to form an array of differences, the final result was in all cases:

Mean of array A = Mean of an array of differences + the number which was subtracted throughout array A to form the array of differences.

Mean using a Guessed Mean

An important property of *all* arrays of data can be stated as follows. The *mean* of any array is equal to the mean of an array of differences, plus the number which was subtracted throughout the original array in order to form the array of differences.

This property is the basis of a method for finding the mean of data. The method reduces the calculations needed to find the mean, especially when there is a large amount of data.

The method involves guessing what the mean of a set of data is likely to be and then using this guess to find the actual mean of the data. A guessed mean is also frequently called either an *assumed mean* or a *working mean*.

To find the actual mean using a guessed mean:

(a) make a sensible guess as to what the actual mean is likely to be.
(b) subtract the guess from every value in the array of data.
(c) find the mean of the array of differences obtained in (b).
(d) add answers to (c) and (a) together i.e. Actual Mean=Mean of Diffs.+Guess.
(e) write down the answer to (d) as the Actual Mean.

Example: Find the mean of the following array.

73 74 77 76 74 76

Either:

(a) guess mean is 76.

(b) array of diffs. is −3, −2, 1, 0, −2, 0.

(c) 6 values
$-3+(-2)+1+0+(-2)+0=-6.$
$-6 \div 6 = -1.$
mean of diffs. = −1.

(d) −1+76=75.

(e) actual mean = 75.

Or:

(a) guess mean is 74.

(b) array of diffs. is −1, 0, 3, 2, 0, 2.

(c) 6 values.
$-1+0+3+2+0+2=6.$
$6 \div 6 = 1.$
mean of diffs. = 1.

(d) 1+74=75.

(e) actual mean = 75.

Or:

(a) guess mean is 73.

(b) array of diffs. is −2, −1, 2, 1, −1, 1.

(c) 6 values.
$-2+(-1)+2+1+(-1)+1=0.$
$0 \div 6 = 0.$
mean of diffs. = 0.

(d) 0+75=75.

(e) actual mean = 75.

Or:

(a) guess mean is 50.

(b) array of diffs. is 23, 24, 27, 26, 24, 26.

(c) 6 values.
$23+24+27+26+24+26=150.$
$150 \div 6 = 25.$
mean of diffs. = 25.

(d) 25+50=75.

(e) actual mean = 75.

Note: All four guesses, although different, give the same actual mean. This shows that any guess will result in the actual mean. It is not possible to make a wrong guess, only a *bad* guess. A sensible guess reduces the amount of calculation needed. A bad guess requires more calculations than a sensible guess.

$$\text{Mean of a set of differences} = \frac{\text{Total of the differences}}{\text{Number of values}}$$

A difference between a value in a set of data and a guessed mean, i.e. (value − guess), is usually denoted by the letter d.

The total of all the differences in a set of data is denoted by Σd.

The number of values in a set of data is denoted by Σf or N.

$$\text{The mean of a set of differences} = \frac{\Sigma d}{\Sigma f} \text{ or } \frac{\Sigma d}{N}$$

To find the mean of large amounts of ungrouped data using a Guessed Mean:

(a) construct a frequency distribution.

(b) make three new columns on the right of the frequency column.

x Value	Tallies	f Freq.	g Guess	$(x-g)$ or d Differences Value – Guess	$f(x-g)$ or fd. Frequency × Differences Frequency × (Value – Guess)

(c) make a sensible guess of what the mean is likely to be.

(d) write down this guess on every line of the first new column, i.e. Guess.

(e) for each line of the distribution, subtract the guess from the number in the original value column.

(f) put the differences obtained from (e) in the second new column, i.e. Value – Guess or Differences d.

(g) for each line of the distribution, multiply the difference in the second new column by the frequency of that difference.

(h) put the answers to (g) in the third new column, i.e. freq. × (Value – Guess) or fd.

(i) add up the third new column. This gives the total of the differences.

(j) add up the frequency column. This gives the number of values in the data.

(k) use the answer to (i), (j) and (c) in the following formula.

$$\text{Actual Mean} = \frac{\text{Total of Diffs.}}{\text{Number of Values}} + \text{Guess}$$

$$\text{or} = \frac{\Sigma fd}{\Sigma f} + \text{Guess}$$

(l) write down the answer to (k) as the actual mean.

Example: Find the mean of the following array.

26 24 25 24 27 25 28 27 29 27 25 27 26 24 24
29 25 25 27 24 25 26 26 28 27 25 28 24 27 26

The frequency distribution gives:

Value	Tallies	Freq.	Guess	Val – Guess	Freq. × (Val – Guess)
24	IIII I	6	27	−3	−18
25	IIII II	7	27	−2	−14
26	IIII	5	27	−1	−5
27	IIII II	7	27	0	0
28	III	3	27	1	3
29	II	2	27	2	4

        30   7 −37

Number of Values   −30

               Total of Diffs.

Total Diffs. $= -30$
Number of Values $= 30$
Guess $= 27$.

$$\text{Actual} = \frac{-30}{30} + 27$$
$$= -1 + 27$$
$$= 26$$

Mean $= 26$

Example: Find the mean of the following array.

```
91  94  94  98  93  95  93  97  95  93
92  94  91  97  96  92  92  96  92  95
```

The frequency distribution gives:

x Value	Tallies	f Freq.	g	d x−g	fd f(x−g)
91	II	2	94	−3	−6
92	IIII	4	94	−2	−8
93	III	3	94	−1	−3
94	III	3	94	0	0
95	III	3	94	1	3
96	II	2	94	2	4
97	II	2	94	3	6
98	I	1	94	4	4

Total Diffs=0
Number of Values=20
Guess=94
Actual mean=$\frac{0}{20}$+94
$\qquad$ =0+94
$\qquad$ =94
Mean $\qquad$ =94

$\Sigma f=20$ — Number of Values

$17-17$ — $\Sigma fd=0$ — Total of diffs.

Example: Find the mean of the following array.

```
0   2  8  9  1  1  2  3  4  9
2   6  3  2  5  2 10  8  4  2
3   7  3 10  5  3  3  6  5  7
```

The frequency distribution gives:

x	Tallies	f	g	d x−g	fd f(x−g)
0	I	1	3	−3	−3
1	II	2	3	−2	−4
2	HHI I	6	3	−1	−6
3	HHI I	6	3	0	0
4	II	2	3	1	2
5	III	3	3	2	6
6	II	2	3	3	6
7	II	2	3	4	8
8	II	2	3	5	10
9	II	2	3	6	12
10	II	2	3	7	14

$\Sigma fd=45$
$\Sigma f=30$
Guess= 3
Actual Mean=$\frac{\Sigma fd}{\Sigma f}$+Guess
$\qquad =\frac{45}{30}+3$
$\qquad =1.5+3$
$\qquad =4.5$
Mean=4.5

$\Sigma f=30$ — Number of Values

$58-13$ — $\Sigma fd=45$ — Total of diffs.

Example: Find the mean of the frequency distribution below.

x	250	252	254	256	258	260	262	264	266	268	270	272	274
f	2	4	4	4	14	20	15	15	10	5	3	3	1

Rewriting the frequency distribution gives:

x	f	g	d	fd
250	2	262	−12	−24
252	4	262	−10	−40
254	4	262	−8	−32
256	4	262	−6	−24
258	14	262	−4	−56
260	20	262	−2	−40
262	15	262	0	0
264	15	262	2	30
266	10	262	4	40
268	5	262	6	30
270	3	262	8	24
272	3	262	10	30
274	1	262	12	12

$\Sigma f = 100$

166−216
$\Sigma fd = -50$

$\Sigma fd = -50$
$\Sigma f = 100$
Guess = 262

$$\text{Actual Mean} = \frac{-50}{100} + 262$$
$$= -0.5 + 262$$
$$= 261.5$$
$$\text{Mean} = 261.5$$

Exercise 10a:
Do these questions using an assumed mean in each case.

1. Find the mean of the following frequency distribution.

Value	10	20	30	40	50
Freq.	2	4	8	12	4

2. Find the mean of the following frequency distribution.

x	10	15	20	25	30	35	40
freq.	2	3	7	10	9	5	4

3. Find the mean of the following frequency distribution.

x	527	528	529	530	531	532
freq.	8	13	4	7	2	6

4. Find the mean of the following frequency distribution.

x	21	22	23	24	25	26	27	28
freq.	1	1	6	13	13	11	4	1

5. Find the mean of the following frequency distribution.

x	92	94	96	98	100	102	104	106	108	110
freq.	1	1	3	5	8	4	2	2	1	3

Mean of Grouped Data using a Guessed Mean

Consider the frequency distribution:

	1–	11–	21–	31–	41–	51–	61–	71–	81–	91–
Freq.	10	20	30	40	50	60	70	80	90	100
Class Interval	1	3	5	8	10	10	5	8	2	2

The mid-values of the class intervals

1–10 11–20 21–30 31–40 41–50 51–60 61–70 71–80 81–90 91–100 are
 5.5   15.5   25.5   35.5   45.5   55.5   65.5   75.5   85.5   95.5

The class size of each interval is 10.
The difference between any two successive mid-values is 10,
   e.g. $55.5 - 45.5 = 10$
         $95.5 - 85.5 = 10$

The difference between any two successive mid-values is the same as the size of each of the class intervals, i.e. 10.
If 55.5 was taken as a guess of the mean of the distribution, then the mid-value 45.5 is 10 below the guess of 55.5.
Similarly 65.5 is 10 above the guess of 55.5.
The difference between 45.5 and the guess is -10.
The difference between 65.5 and the guess is $+10$.
The mid-values of the intervals are

 5.5   15.5   25.5   35.5   45.5   55.5   65.5   75.5   85.5 95.5

The differences of these mid-values from a guessed mean of 55.5 are
 -50 -40 -30 -20 -10 0   10   20   30   40.

The differences in the above example are multiples of the class size 10.

    $10 = 1 \times 10$ $40 = 4 \times 10$ $-20 = -2 \times 10$ $-50 = -5 \times 10$
    $20 = 2 \times 10$ $0 = 0 \times 10$ $-30 = -3 \times 10$
    $30 = 3 \times 10$ $-10 = -1 \times 10$ $-40 = -4 \times 10$

Similarly for

Intervals	74–77	78–81	82–85	86–89	90–93	94–97
mid-values	75.5	79.5	83.5	87.5	91.5	95.5
diffs. from 87.5	-12	-8	-4	0	4	8
multiples of class size	-3×4	-2×4	-1×4	0×4	1×4	2×4

Class size=4
diffs. between
successive
intervals
is 4.

The differences between the mid-values of a distribution and a guessed mean can be used in the calculation of the mean.

When data has been grouped, the *actual* mean of the data can only be estimated.

To find an estimate of the actual mean of grouped data using a guessed mean:

(a) construct a frequency distribution.
(b) make four new columns on the right of the frequency column.

Interval	Tallies	*f* Freq.	*X* Mid. Value	*G* Guess	(X−G) or D Differences Mid. Val.−Guess	f(X−G) or fD Freq.×Differences Freq.×(Mid. Val.−Guess)

(c) find the mid-value of each interval and put them in the first new column, i.e. Mid.value.
(d) make a sensible guess of what the mean is likely to be but ensure that it is one of the *mid-values* from (c).
(e) write down this guess on every line of the second new column, i.e. Guess.
(f) for each line of the distribution subtract the guess from the mid-value of that line.
(g) put the differences obtained from (f) in the third new column, i.e. Mid.Value−Guess or Differences D.
(h) for each line of the distribution multiply the difference in the third new column by the frequency of that difference.
(i) put the answers to (h) in the fourth new column, i.e. freq.×(Mid. Value−Guess) or fD.
(j) add up the fourth new column. This gives the total of the differences.
(k) add up the frequency column. This gives the number of values in the data.
(l) use the answer to (j), (k), and (d) in the following formula.

$$\text{Actual Mean} = \frac{\text{Total of Diffs.}}{\text{Number of Values}} + \text{Guess}$$

$$Or=\frac{\Sigma fD}{\Sigma f}+\text{Guess}$$

(*m*) write down the answer to (*l*) as an *estimate* of the actual mean.

Example: Find the mean of the following array by grouping into class intervals of 74–77, 78–81, 82–85 etc. and then using a guessed mean.

86　82　90　88　95　83　96　75　83　86　91　87
91　79　87　92　84　88　86　85　93　80　87　89

The frequency distribution gives:

Interval	Tallies	f Freq.	Mid. Val.	Guess	MV−G	f(MV−G)
74–77	I	1	75.5	87.5	−12	−12
78–81	II	2	79.5	87.5	−8	−16
82–85	ӇӇ	5	83.5	87.5	−4	−20
86–89	ӇӇ IIII	9	87.5	87.5	0	0
90–93	ӇӇ	5	91.5	87.5	4	20
94–97	II	2	95.5	87.5	8	16

24
Number of Values

36−48
−12
Total of diffs.

Total of Diffs.= −12
Number of values=24
Guess=87.5

Actual Mean= $\dfrac{-12}{24}+87.5$

　　　　　=−0.5+87.5
　　　　　=87
　　Mean=87

Example: Find the mean of the following array by grouping into class intervals of 0–4, 5–9, 10–14 etc. and then using a guessed mean.

27　8　22　28　7　23　32　37　26　48
13　33　27　12　14　34　24　18　43　17
41　46　21　21　47　10　42　29　30　2
5　38　26　35　21　31　36　25　15　40

The frequency distribution gives:

Interval	Tallies	f Freq.	X MV	G	$(X-G)$ or D $MV-G$	fD $f(MV-G)$
0–4	I	1	2	27	−25	−25
5–9	III	3	7	27	−20	−60
10–14	IIII	4	12	27	−15	−60
15–19	III	3	17	27	−10	−30
20–24	HHf I	6	22	27	−5	−30
25–29	HHf II	7	27	27	0	0
30–34	HHf	5	32	27	5	25
35–39	IIII	4	37	27	10	40
40–44	IIII	4	42	27	15	60
45–49	III	3	47	27	20	60

Total Diffs.$= -20$
Number of
Values$=40$
Guess$=27$

$$\text{Actual Mean} = \frac{-20}{40}+27$$
$$= -0.5+27$$
$$=26.5$$
Mean $=26.5$

$\Sigma f=40$
Number of Values

$185-205$
$\Sigma fD = -20$
Total of Diffs.

Example: Find the mean of the frequency distribution below by using a guessed mean.

Interval	0– 9	10– 19	20– 29	30– 39	40– 49	50– 59	60– 69	70– 79	80– 89	90– 99
Freq.	2	4	6	14	22	16	7	5	3	1

Rewriting the frequency distribution gives:

Interval	f	X MV	G	D $X-G$	fD
0–9	2	$4\frac{1}{2}$	$44\frac{1}{2}$	−40	−80
10–19	4	$14\frac{1}{2}$	$44\frac{1}{2}$	−30	−120
20–29	6	$24\frac{1}{2}$	$44\frac{1}{2}$	−20	−120
30–39	14	$34\frac{1}{2}$	$44\frac{1}{2}$	−10	−140
40–49	22	$44\frac{1}{2}$	$44\frac{1}{2}$	0	0
50–59	16	$54\frac{1}{2}$	$44\frac{1}{2}$	10	160
60–69	7	$64\frac{1}{2}$	$44\frac{1}{2}$	20	140
70–79	5	$74\frac{1}{2}$	$44\frac{1}{2}$	30	150
80–89	3	$84\frac{1}{2}$	$44\frac{1}{2}$	40	120
90–99	1	$94\frac{1}{2}$	$44\frac{1}{2}$	50	50

$\Sigma fD =160$
$\Sigma f=80$
Guess$=44\frac{1}{2}=44.5$
$$\text{Actual Mean} = \frac{160}{80}+44.5$$
$$=2+44.5$$
$$=46.5$$
Mean $=46.5$

$\Sigma f=80$
Number of Values

$620-460$
$\Sigma fD =160$
Total Diffs.

Example: Find the mean of the following array by grouping into class intervals of 1–10, 11–20 etc. and then using a guessed mean.

51	71	23	46	53	63	56	95	43	57	91	36	72	69	31
81	48	14	5	51	53	61	56	73	68	47	32	63	49	41
42	51	66	53	45	57	38	40	76	74	56	52	54	27	50
62	17	70	85	57	82	29	42	52	77	25	62	67	68	89

The frequency distribution gives:

Interval	Tallies	f	X	G	D	fD
1–10	I	1	$5\frac{1}{2}$	$55\frac{1}{2}$	−50	−50
11–20	II	2	$15\frac{1}{2}$	$55\frac{1}{2}$	−40	−80
21–30	IIII	4	$25\frac{1}{2}$	$55\frac{1}{2}$	−30	−120
31–40	HHT	5	$35\frac{1}{2}$	$55\frac{1}{2}$	−20	−100
41–50	HHT HHT	10	$45\frac{1}{2}$	$55\frac{1}{2}$	−10	−100
51–60	HHT HHT HHT	15	$55\frac{1}{2}$	$55\frac{1}{2}$	0	0
61–70	HHT HHT I	11	$65\frac{1}{2}$	$55\frac{1}{2}$	10	110
71–80	HHT I	6	$75\frac{1}{2}$	$55\frac{1}{2}$	20	120
81–90	IIII	4	$85\frac{1}{2}$	$55\frac{1}{2}$	30	120
91–100	II	2	$95\frac{1}{2}$	$55\frac{1}{2}$	40	80

$\Sigma f=60$

430–450
$\Sigma fD= -20$

$\Sigma fD= -20$
$\Sigma f=60$
Guess$=55\frac{1}{2}=55.5$
$\frac{\text{Actual}}{\text{Mean}} =\frac{-20}{60}+55.5$
$=-0.3\dot{3}+55.5$
$=55.17$ to 2
Decimal Places.
Mean $=55.17$

The class intervals of a distribution need not all be of equal class size. Unequal class intervals do not cause any problems, as can be seen from the working in the table below. The mid-values of the intervals are still used.

Class	f	X	G	D	fD	
1–20	3	10.5	55.5	−45	−135	
21–30	4	25.5	55.5	−30	−120	
31–40	8	35.5	55.5	−20	−160	
41–50	10	45.5	55.5	−10	−100	
51–60	10	55.5	55.5	0	0	
61–70	6	65.5	55.5	10	60	
71–80	6	75.5	55.5	20	120	
81–100	3	90.5	55.5	35	105	

$\Sigma f = 50$

$$285 - 515$$
$$\Sigma fD = -230$$

For the mean
$$\Sigma fD = -230$$
$$\Sigma f = 50$$
$$\text{Guess} = 55.5$$
$$\text{Mean} = \frac{\Sigma fD}{\Sigma f} + G$$
$$= \frac{-230}{50} + 55.5$$
$$= -4.6 + 55.5$$
$$\text{Mean} = 50.9$$

Exercise 10b:
Do each of the following questions using an assumed mean.

1. Find the mean of the following frequency distribution.

Class interval	1–20	21–40	41–60	61–80	81–100
Freq.	7	16	23	10	4

2. Find the mean of the following frequency distribution.

Class interval	28–32	33–37	38–42	43–47	48–52	53–57
Freq.	1	4	13	22	16	4

3. Find the mean of the following frequency distribution.

Class interval	1–10	11–20	21–30	31–40	41–50	51–60	61–70	71–80	81–90	91–100
Freq.	3	4	8	14	38	48	27	6	1	1

4. Find the mean of the following frequency distribution.

Class interval	1–100	101–200	201–300	301–400	401–500	501–600	601–700
Freq.	1	11	14	25	15	10	4

5. Find the mean of the following frequency distribution.

Class interval	1– 5	6– 10	11– 15	16– 20	21– 25	26– 30	31– 35	36– 40	41– 45	46– 50
Freq.	4	7	12	13	20	39	54	34	10	7

Standard Deviation using a Guessed Mean

It can be proved that the *variance* of a set of data is given by the formula

$$\text{Variance} = \frac{\Sigma fd^2}{\Sigma f} - \left(\frac{\Sigma fd}{\Sigma f}\right)^2 \text{ or } \frac{\Sigma fD^2}{\Sigma f} - \left(\frac{\Sigma fD}{\Sigma f}\right)^2$$

where d = difference of the values in a set of ungrouped data from a guessed
mean.

D = difference of the mid-values of a set of grouped data from a
guessed mean.

To find the standard deviation using a guessed mean:

(a) find the mean using a guessed mean as described on pages 215
and 219.

(b) make one new column on the right of the fd or fD column.

fd^2 Frequency × differences² frequency × (value−guess)²	or	fD^2 frequency × differences² frequency × (Mid. Val. − Guess) ²

(c) for each line of the distribution multiply the number in the differences column, d or D, by the number in the fd or fD column.

(d) put the answers to (c) in the new column, i.e. frequency × differ-
ences², fd^2 or fD^2.

(e) add up the new column. This gives the total of the column, i.e. Σfd^2
or ΣfD^2.

(f) use the answer to (e) and the values obtained for Σf and Σfd or ΣfD
obtained during step (a) in the following formula.

$$\text{Variance} = \frac{\Sigma fd^2}{\Sigma f} - \left(\frac{\Sigma fd}{\Sigma f}\right)^2 \text{ or } \frac{\Sigma fD^2}{\Sigma f} - \left(\frac{\Sigma fD}{\Sigma f}\right)^2$$

(g) find the square root of the answer to (f).

(h) write down the answer to (g) as the standard deviation.

Example: Find the mean and standard deviation of the frequency distribution below.

x	0	1	2	3	4	5	6	7	8	9	10
f	2	4	8	11	13	19	18	12	10	2	1

x	f	g	d	fd	fd²
0	2	6	−6	−12	72
1	4	6	−5	−20	100
2	8	6	−4	−32	128
3	11	6	−3	−33	99
4	13	6	−2	−26	52
5	19	6	−1	−19	19
6	18	6	0	0	0
7	12	6	1	12	12
8	10	6	2	20	40
9	2	6	3	6	18
10	1	6	4	4	16

$\Sigma f = 100$

$42 - 142$ 556

$\Sigma fd = -100$ Σfd^2

For Mean

$\Sigma fd = -100$

$\Sigma f = 100$

$$\text{Mean} = \frac{\Sigma fd}{\Sigma f} + g$$

$$= \frac{-100}{100} + 6$$

$$= -1 + 6$$

$$\text{Mean} = 5$$

For standard deviation

$$\text{Variance} = \frac{\Sigma fd^2}{\Sigma f} - \left(\frac{\Sigma fd}{\Sigma f}\right)^2$$

$$= \frac{556}{100} - \left(\frac{-100}{100}\right)^2 = 5.56 - (-1)^2 = 5.56 - (1)$$

$$= 5.56 - 1 = 4.56$$

Standard deviation $= \sqrt{\text{Variance}} = \sqrt{4.56} = 2.135 = 2.14$ to 2 D.P.

Example: Find the mean and standard deviation of the frequency distribution below.

Class	1– 7	8– 14	15– 21	22– 28	29– 35	36– 42	43– 49	50– 56	57– 63	64– 70
Freq.	2	2	3	4	14	10	6	5	3	1

Rewriting the frequency distribution gives:

Class	f	X	G	D	fD	fD^2
1–7	2	4	32	−28	−56	1568
8–14	2	11	32	−121	−42	882
15–21	3	18	32	−14	−42	588
22–28	4	25	32	−7	−28	196
29–35	14	32	32	0	0	0
36–42	10	39	32	7	70	490
43–49	6	46	32	14	84	1176
50–56	5	53	32	21	105	2205
57–63	3	60	32	28	84	2352
64–70	1	67	32	35	35	1225

For the mean
$\Sigma fD = 210$
$\Sigma f = 50$
$G = 32$

$$\text{Mean} = \frac{\Sigma fD}{\Sigma f} + G$$

$$= \frac{210}{50} + 32$$

$$= 4.2 + 32$$

$$\text{Mean} = 36.2$$

$\Sigma f = 50$

$378 - 168 \quad 10682$
$\Sigma fD = 210 \quad \Sigma fD^2$

For the standard deviation

$$\text{Variance} = \frac{\Sigma fD^2}{\Sigma f} - \left(\frac{\Sigma fD}{\Sigma f}\right)^2$$

$$= \frac{10682}{50} - \left(\frac{210}{50}\right)^2$$

$$= 213.64 - (4.2)^2$$

$$= 213.64 - 17.64$$

$$= 196$$

$$\text{Standard deviation} = \sqrt{\text{Variance}}$$

$$= \sqrt{196} = 14$$

Mean and Standard Deviation

The mean and standard deviation of grouped data using *multipliers* of the class size from a guessed mean.

It was shown on page 218 that the difference between the mid-vlaues of a distribution and a guessed mean are multiples of the class size of the distribution.

E.g. the differences −50 −40 −30 −20 −10 0 10 20 30 40 were multiples of the class size 10.

The differences −12 −8 −4 0 4 8 were multiples of the class size 4.

The differences of a distribution can be expressed as:

a *multiplier* × class size.

Differences	−50	−40	−30	−20	−10	0
Multiples of 10	−5×10	−4×10	−3×10	−2×10	−1×10	0×10
Multiplier	−5	−4	−3	−2	−1	0

Differences	10	20	30	40
Multiples of 10	1×10	2×10	3×10	4×10
Multiplier	1	2	3	4

Similarly:

Differences	−12	−8	−4	0	4	8
Multiples of 4	−3×4	−2×4	−1×4	0×4	1×4	2×4
Multiplier	−3	−2	−1	0	1	2

Or:

Differences	−28	−21	−14	−7	0	7
Multiples of 7	−4×7	−3×7	−2×7	−1×7	0×7	1×7
Multiplier	−4	−3	−2	−1	0	1

Differences	14	21	28	35
Multiples of 7	2×7	3×7	4×7	5×7
Multiplier	2	3	4	5

The multipliers need not be whole numbers,
e.g. in the case of unequal class intervals.

Interval	1–20	21–30	31–40	41–50	51–60
Mid-value	10.5	25.5	35.5	45.5	55.5
Diffs. from 45.5	−35	−20	−10	0	10
Multiples of 10	−3.5×10	−2×10	−1×10	0×10	1×10
Multiplier	−3.5	−2	−1	0	1

Interval	61–70	71–80	81–100
Mid-value	65.5	75.5	90.5
Diffs. from 45.5	20	30	45
Multiples of 10	2×10	3×10	4.5×10
Multiplier	2	3	4.5

The calculations involved in finding the mean and standard deviation using a guessed mean, can be simplified using *multipliers* of the class size in place of the difference.

The column D can be replaced by D_m, i.e. differences expressed
Similarly fD can be replaced by fD_m as multipliers.
fD^2 can be replaced by fD_m^2
ΣfD can be replaced by ΣfD_m
ΣfD^2 can be replaced by ΣfD_m^2

The steps involved in the calculation of the mean and standard deviation are exactly the same as those described on pages 215, 219 and 224 but multipliers are used instead of differences.

The formula used to find the mean is however adjusted to:

$$\text{Mean} = S \times \frac{\Sigma fD_m}{\Sigma f} + Guess$$

where S is the class size of the intervals.
The formula for the variance is replaced by

$$\text{Variance of multipliers} = \frac{\Sigma fD_m^2}{\Sigma f} - \left(\frac{\Sigma fD_m}{\Sigma f}\right)^2$$

and standard deviation $= S \times \sqrt{\text{Variance of multipliers}}$ where S is the class size of the intervals.

Example: Find the mean and standard deviation of the frequency distribution shown below.

Class	0–9	10–19	20–29	30–39	40–49	50–59	60–69	70–79	80–89	90–99
Freq.	2	4	5	10	22	20	8	5	3	1

Rewriting the frequency distribution and using multipliers gives:

Class	f	X	G	D_m	fD_m	fD_m^2
0–9	2	4.5	44.5	−4	−8	32
10–19	4	14.5	44.5	−3	−12	36
20–29	5	24.5	44.5	−2	−10	20
30–39	10	34.5	44.5	−1	−10	10
40–49	22	44.5	44.5	0	0	0
50–59	20	54.5	44.5	1	20	20
60–69	8	64.5	44.5	2	16	32
70–79	5	74.5	44.5	3	15	45
80–89	3	84.5	44.5	4	12	48
90–99	1	94.5	44.5	5	5	25

$\Sigma f = 80$ 68−40   268
$\Sigma fD_m = 28$ ΣfD_m^2

For the mean

Class Size $= S = 10$

$\Sigma fD_m = 28$

$\Sigma f = 80$

$G = 44.5$

$$\text{Mean} = S \times \frac{\Sigma fD_m}{\Sigma f} + Guess$$

$$= 10 \times \frac{28}{80} + 44.5$$

$$= \frac{280}{80} + 44.5$$

$$= 3.5 + 44.5$$

$$\text{Mean} = 48.0$$

For the standard deviation

Variance of multipliers
$$= \frac{\Sigma fD_m^2}{\Sigma f} - \left(\frac{\Sigma fD_m}{\Sigma f}\right)^2$$

$$= \frac{268}{80} - \left(\frac{28}{80}\right)^2$$

$$= 3.35 - (0.35)^2$$

$$= 3.35 - 0.1225 = 3.2275$$

Standard deviation $= S \times \sqrt{Variance\ of\ multipliers}$

$$= 10 \times \sqrt{3.2275}$$

$$= 10 \times 1.796$$

$$= 17.96$$

Example: Find the mean and standard deviation of the frequency distribution shown below.

Class	1–7	8–14	15–21	22–28	29–35	36–42	43–49	50–56	57–63	64–70
Freq.	2	2	3	4	14	10	6	5	3	1

Rewriting the frequency distribution and using multipliers gives:

Class	f	X	G	D_m	fD_m	fD_m^2
1–7	2	4	32	−4	−8	32
8–14	2	11	32	−3	−6	18
15–21	3	18	32	−2	−6	12
22–28	4	25	32	−1	−4	4
29–35	14	32	32	0	0	0
36–42	10	39	32	1	10	10
43–49	6	46	32	2	12	24
50–56	5	53	32	3	15	45
57–63	3	60	32	4	12	48
64–70	1	67	32	5	5	25

$$\Sigma f = 50 \qquad\qquad \begin{array}{cc} 54-24 & 218 \end{array}$$
$$\Sigma fD_m = 30 \quad \Sigma fD_m^2$$

For the mean
Class Size = $S = 7$
$\Sigma fD_m = 30$
$\Sigma f = 50$
$G = 32$

$$\text{Mean} = S \times \frac{\Sigma fD_m}{\Sigma f} + G$$

$$= 7 \times \frac{30}{50} + 32$$

$$= \frac{210}{50} + 32$$

$$= 4.2 + 32$$

Mean $= 36.2$

For the standard deviation

$$\begin{array}{l}\text{Variance}\\ \text{of multipliers}\end{array} = \frac{\Sigma fD_m^2}{\Sigma f} - \left(\frac{\Sigma fD_m}{\Sigma f}\right)^2$$

$$= \frac{218}{50} - \left(\frac{30}{50}\right)^2$$

$$= 4.36 - (0.6)^2$$

$$= 4.36 - 0.36$$

$$= 4.00$$

$$\text{Standard deviation} = S \times \sqrt{\text{Variance of multipliers}}$$

$$= 7 \times \sqrt{4}$$

$$= 7 \times 2$$

$$= 14$$

The working in the table below shows the calculation of the mean and standard deviation when unequal class sizes are present.

Class	f	X	G	D_m	fD_m	fD_m^2
1–20	3	10.5	55.5	−4.5	−13.5	60.75
21–30	4	25.5	55.5	−3	−12	36
31–40	8	35.5	55.5	−2	−16	32
41–50	10	45.5	55.5	−1	−10	10
51–60	10	55.5	55.5	0	0	0
61–70	6	65.5	55.5	1	6	6
71–80	6	75.5	55.5	2	12	24
81–100	3	90.5	55.5	3.5	10.5	36.75

$$\Sigma f = 50 \qquad\qquad \begin{matrix} 28.5-51.5 & 205.5 \\ \Sigma fD_m = -23 & \Sigma fD_m^2 \end{matrix}$$

For the mean
Class Size taken as 10 = S

$$\Sigma fD_m = -23$$
$$\Sigma f = 50$$
$$G = 55.5$$
$$\text{Mean} = S \times \frac{\Sigma fD_m}{\Sigma f} + G$$
$$= 10 \times \frac{(-23)}{50} + 55.5$$
$$= \frac{-230}{50} + 55.5$$
$$= -4.6 + 55.5$$
$$\text{Mean} = 50.9$$

For standard deviation

$$\begin{aligned} \text{Variance of multipliers} &= \frac{\Sigma fD_m^2}{\Sigma f} - \left(\frac{\Sigma fD_m}{\Sigma f}\right)^2 \\ &= \frac{205.5}{50} - \left(\frac{-23}{50}\right)^2 \\ &= 4.11 - (-0.46)^2 \\ &= 4.11 - 0.2116 = 3.8984 \end{aligned}$$
$$\begin{aligned} \text{Standard deviation} &= S \times \sqrt{\text{Variance of multipliers}} \\ &= 10 \times \sqrt{3.8984} \\ &= 10 \times 1.974 \\ &= 19.74 \end{aligned}$$

Exercise 10c:
Do the following questions using an assumed mean.

1. Find the standard deviation of the frequency distribution of:

(*a*) question 1 of Exercise 10a.
(*b*) question 2 of Exercise 10a.
(*c*) question 3 of Exercise 10a.
(*d*) question 4 of Exercise 10a.
(*e*) question 5 of Exercise 10a.

2. Find the standard deviation of the frequency distribution of:

(*a*) question 1 of Exercise 10b.
(*b*) question 2 of Exercise 10b.
(*c*) question 3 of Exercise 10b.
(*d*) question 4 of Exercise 10b.
(*e*) question 5 of Exercise 10b.

Miscellaneous Exercises 10

1. (*a*) Over the first part of an athletics season a young athlete keeps a check on the times he records when running 100 m. The results are shown in the table.

Time in seconds	11.7	11.8	11.9	12.0	12.1	12.2
Number of runs	4	6	7	6	4	3

Calculate the arithmetic mean for the frequency distribution using an assumed mean of 12.0 seconds.

(*b*) The marks obtained by 50 pupils in an end of term examination are given in the table.

Marks	Number of pupils	Mid. Interval Mark.
0 to 30	15	15
31 to 50	22	40.5
51 to 70	8	60.5
71 to 100	5	85.5

Calculate the arithmetic mean for this distribution.   NW78

2. (*a*) The following is a frequency table for marks obtained by 60 children in a mock C.S.E. examination.

No. of Marks	Frequency
0–9	2
10–19	4
20–29	4
30–39	14
40–49	23
50–59	9
60–69	3
70–79	1

Calculate the arithmetic mean for the frequency distribution, using an assumed mean in the class interval 40–49 marks. Use the formula

$$M = am \pm \frac{\Sigma fd_i}{N}$$

or any other method using an assumed mean.

(*b*) Find the range, median and mean for:
(i) 5, 6, 7, 11, 11, 12, 15, 17;
(ii) 15, 17, 20, 21, 21, 22, 24, 24, 26, 27, 30.   NW76

3. The members of two forms obtained the following results in a mathematics test marked out of a possible total of 80 marks.

Results:

22	76	46	54	56	41
43	17	44	51	13	15
46	1	68	48	21	27
34	29	41	26	16	25
50	51	22	26	54	19
65	43	25	41	70	37
68	35	42	25	68	47
17	38	10	57	33	31
15	71	26	50	54	29
74	54	35	67	59	61

(*a*) Obtain a frequency distribution using a class interval of 5 marks arranged in *ascending* order e.g., 1–5, 6–10, 11–15 etc.

(*b*) Use your frequency distribution to calculate the Arithmetic Mean using

$$M = Am \pm \frac{\Sigma fd_i}{N}$$

(or any other method that you know) and an Assumed Mean in the class interval 36–40 marks.

  N.B.—1. Part (*b*) depends upon an accurate frequency distribution so take great care to check your frequencies.
  2. Follow the instruction as to your choice of Assumed Mean.
  NW74

4. The following table gives an analysis by unladen weight of Road Goods Vehicles in Great Britain in 1975.

Unladen weight (tons)	*No. of Vehicles (10 thousands)*
Not over 1 ton	63
1 ton but not over 2 tons	55
2 tons but not over 3 tons	13
3 tons but not over 5 tons	16
5 tons but not over 8 tons	13
8 tons but not over 15 tons	10

Source: Dept. of the Environment.

(*a*) Write down the mid-point of each class interval.
(*b*) Using an assumed mean of 2.5 tons, calculate, correct to one place of decimals, the mean unladen weight of the vehicles and the standard deviation.   EA78

5. Find the mean and standard deviation of the frequency distribution below.

Class	1–10	11–20	21–30	31–40	41–50	51–60	61–70	71–80	81–90	91–100
Freq.	2	5	20	38	55	46	22	8	3	1

6. Find the mean and standard deviation of the frequency distribution below.

Class	31–35	36–40	41–45	46–50	51–55	56–60	61–65	66–70
Freq.	1	2	6	13	38	26	10	4

7. Find the mean and standard deviation of the frequency distribution below.

Class	0– 4	5– 9	10– 14	15– 19	20– 24	25– 29	30– 34	35– 39	40– 44
Freq.	1	2	5	16	50	26	17	2	1

8. The table below is a frequency distribution of the life in hours, to the nearest hour, of 1000 electric lamps.

Life (in hours)	Frequency
600–699	30
700–799	115
800–899	150
900–999	190
1000–1099	170
1100–1199	150
1200–1299	120
1300–1399	55
1400–1499	20

Taking care to get the class boundaries correct, calculate:

(*a*) the mean life
(*b*) the standard deviation. WM78

9. The heights of 100 men are measured to the nearest inch, and the following frequency distribution is obtained:

Height (in inches)	60–	62–	64–	66–	68–	70–	72–	74–75
Frequency	3	8	11	19	32	14	10	3

Taking care to get the class boundaries correct,

calculate (i) the mean height,
(ii) the standard deviation. WM75

10. A village has 900 inhabitants. Their ages are distributed according to the following frequency distribution:

Age in years	0–	10–	20–	30–	40–	50–	60–	70–	80–	90–100
Frequency	159	137	114	125	122	118	73	40	9	3

calculate (i) the arithmetic mean of these ages
　　　　　(ii) the standard deviation.　　　　　　　　WM76

11. The weights of 120 men are measured to the nearest lb, and the following frequency distribution is obtained:

Weight (in lb)	110–	130–	150–	170–	190–	210–229
Frequency	2	20	43	34	16	5

Taking care to get the class boundaries correct,
　calculate (i) the mean weight,
　　　　　(ii) the standard deviation.　　　　　　　WM77

11 WEIGHTED AVERAGES

The average of a set of data is a value chosen to be representative of that set of data.
The mean is one type of average.

The mean is defined as: $\dfrac{\Sigma fx}{\Sigma f} = \dfrac{\text{Total of Data}}{\text{Number of values of data}}$

Situations arise where to take a straightforward mean of the values of a set of items would lead to a distorted view of the average value of the items.
E.g. a housewife buys the following items at the greengrocers each week.

potatoes at 4p per lb   carrots at 5p per lb   onions at 10p per lb
turnips at 11p per lb   tomatoes at 44p per lb   parsnips at 10p per lb

Considering the mean cost per lb as $\dfrac{\text{Total of the costs per lb}}{\text{Number of items in the list}}$

we obtain mean cost per lb $= \dfrac{4 + 5 + 10 + 11 + 44 + 10}{6}$

$$= 84 \div 6 = 14\text{p}.$$

This is obviously a distorted view of the mean cost per lb, since only one item, tomatoes, has a cost of more than 14p per lb.
The mean in this case is completely unrepresentative of the items in the list.
To obtain a more accurate picture of this type of situation we must take into account not only how much each item costs per lb, but also what quantity, i.e. *weight*, the housewife buys of each item. If we take into account the *weights* bought, we can redefine the mean cost per lb as:

$$\text{Mean cost per lb} = \dfrac{\text{Total cost of the food}}{\text{Number of lbs of food bought}}$$

The housewife found that each week she bought:

10 lb potatoes   5 lb carrots   2 lb onions
1 lb turnips   1 lb tomatoes   1 lb parsnips

The total cost of a week's shopping is given by:

potatoes $4 \times 10 = 40$
carrots $5 \times 5 = 25$
onions $10 \times 2 = 20$
turnips $11 \times 1 = 11$
tomatoes $44 \times 1 = 44$
parsnips $10 \times 1 = \underline{10}$
            $\quad\quad\quad 150\text{p} = \text{Total Cost}$

Since the total cost of the food = 150p
and the number of lbs of food = 20 lb

we have mean cost per lb = $\dfrac{150}{20}$ = $7\frac{1}{2}$p.

This is a much more realistic picture of the average price per lb of the food. The above is an example of a *weighted mean*, where the mean is based not only on the values of the items, but also on the weights of each item. The working for the above example could be laid out as follows:

	c Cost per lb	w Weight	wc Cost × weight
Potatoes	4	10	40
Carrots	5	5	25
Onions	10	2	20
Turnips	11	1	11
Tomatoes	44	1	44
Parsnips	10	1	10

Number lbs = Σw = 20
Total cost = Σwc = 150

$$\begin{array}{r} 7.5 \\ 20\overline{)150} \end{array}$$

$\dfrac{\text{Total Cost}}{\text{Number lbs}} = \dfrac{\Sigma wc}{\Sigma w} = 7\frac{1}{2}$

Mean = $7\frac{1}{2}$p per lb

Σw = 20 Σwc = 150
Number lbs of food   Total cost

It should be noted that this is exactly the same working as that for an ordinary mean as described on page 95, except that:

(*a*) the value column has been replaced by a cost column c.
(*b*) the freq. column f has been replaced by a weight column w.
(*c*) the freq. × value column fx has been replaced by a weight × cost column wc.

If in our weighted mean example we had used x instead of c, the only difference would have been the use of w in the place of f.

To work out a weighted mean:

(*a*) proceed as for an ordinary mean but use *weights* in the place of frequencies.
(*b*) use the following formula.

$$\text{Weighted Mean} = \frac{\text{Total of Data taking weights into account}}{\text{Total of weights}} = \frac{\Sigma wx}{\Sigma w}$$

The idea of a weighted mean can be applied to other situations where weights as such are not involved.

Throughout all walks of life there are always some people who are thought to be more important than others. The more important a person is thought to be, the more notice other people are likely to take of that person.

Similarly, in any list of items there will be some items which are considered to be of more importance than others.

It is often felt that the differences in the levels of importance of the various items should be taken into account in any results worked out from a list of items.

When considering a list of items, the level of importance of each item can be 'weighed' against the levels of importance of all the other items. Each item can then be given an importance rating or weighting.

The greater the importance of an item, the higher is its weighting.

Using this idea of weightings, weighted means of data can be calculated.

E.g. a pupil studies geography, French, music, maths, English, art, woodwork and science.

The headmaster of the school might decide that each subject is not of equal importance, but that some are more important than others.

He could decide that the two most important subjects are English and maths, with the next two most important subjects being French and science, followed by geography and finally equally least important art, woodwork and music.

The headmaster would probably give importance ratings or weights to the subjects as follows: English and maths 4, French and science 3, geography 2, art, woodwork and music 1.

Example: Our pupil obtained the following percentages in the end of year exams.

geography 50   French 45   music 90   maths 40
English 60   art 70   woodwork 30 science 55

- (a) What was his ordinary mean percentage?
- (b) What was his weighted mean percentage using the weights given to each subject by the headmaster?

Ordinary mean = $(50 + 45 + 90 + 40 + 60 + 70 + 30 + 55) \div 8$
$= 440 \div 8$
$= 55\%$

For the weighted mean:

	x Percentage	w Weight	wx Weight × Percentage
Geography	50	2	100
French	45	3	135
Music	90	1	90
Maths	40	4	160
English	60	4	240
Art	70	1	70
Woodwork	30	1	30
Science	55	3	165

$\Sigma w = 19$ $\Sigma wx = 990$
Total   Total Marks
Weights   with weighting

Total marks with weighting $= \Sigma wx = 990$
Total weights $= \Sigma w = 19$

$$19\overline{)990}\ \ 52.1$$

Weighted Mean $= \dfrac{\text{Total marks}}{\text{Total weights}}$

$= \dfrac{\Sigma wx}{\Sigma w} = \dfrac{990}{19} = 52.1\%$

When marks have been weighted, our pupil has a slightly lower average mark than before weighting the marks.

Example: Manganin, an alloy of three metals, consists of 84% copper, 4% nickel and 12% manganese. Find the overall density of manganin if the densities of copper, nickel and manganese are $8.96\,\text{g cm}^{-3}$, $8.9\,\text{g cm}^{-3}$ and $7.44\,\text{g cm}^{-3}$ respectively.
To find the overall density of manganin we must find a weighted mean of the densities of copper, nickel and manganese. The weighting to be used with copper is 84, with nickel 4, and with manganese 12, since these are the proportions which make up the manganin.

	x Density	w Weight	wx Weight × Density
Copper	8.96	84	752.64
Nickel	8.9	4	35.60
Manganese	7.44	12	89.28

$\Sigma w = 100$ $\Sigma wx = 877.52$

$\Sigma wx = 877.52$
$\Sigma w = 100$

$$100\overline{)877.52}\ \ 8.7752$$

Weighted Mean = 8.78 to 2D.P.

Overall density of manganin $= 8.78\,\text{g cm}^{-3}$

When working out weighted means:
Both the *units* of an item and that item's *importance* or *weighting* must be taken into account.
The methods of working out means, using a guessed mean, can be used equally well for finding weighted means.

Example: Find the weighted average mark of a pupil taking a mathematics exam which has eight parts to it. Each part of the exam is marked out of 100, but the different parts are given weightings, dependent on the difficulty of that part of the exam. The marks obtained by the pupil and the weightings of each part are shown on the table below.

Part	A	B	C	D	E	F	G	H
Mark	40	60	55	70	80	45	60	50
Weight	9	5	10	7	1	8	6	4

This gives:

Part	x	w	g	d x-g	wd $w(x$-$g)$
A	40	9	55	-15	-135
B	60	5	55	5	25
C	55	10	55	0	0
D	70	7	55	15	105
E	80	1	55	25	25
F	45	8	55	-10	-80
G	60	6	55	5	30
H	50	4	55	-5	-20

$\Sigma w = 50$

185–235
$\Sigma wd = -50$

Guess $= -55$
$\Sigma wd = -50$
$\Sigma w = 50$
$$\frac{\Sigma wd}{\Sigma w} = \frac{-50}{50} = -1$$
Mean $= -1 + 55$
$= 54$

Weighted mean mark = 54.

Price Relatives

The price of an item tends to rise year by year.
E.g. The price of commodity A is shown for various years in the table below.

Year	1976	1977	1978	1979	1980
Price £'s	8	10	12	15	16

The price of an item in any year can be expressed as a percentage of the price in a given *base* year.
E.g. The prices for 1976, 1977, 1979 and 1980 can be expressed as percentages of the price in 1978.

The price for 1976 is $\frac{8}{12} \times 100 = 66.6\%$ of the price in 1978.

for 1977 is $\frac{10}{12} \times 100 = 83.\dot{3}\%$ of the price in 1978.

for 1979 is $\frac{15}{12} \times 100 = 125\%$ of the price in 1978.

for 1980 is $\frac{16}{12} \times 100 = 133.\dot{3}\%$ of the price in 1978.

The price of a commodity in any year can be expressed as a percentage *relative* (in relation) to the price in some given *base* year.

In our example the prices of commodity A in various years have been calculated as percentages relative to the price in 1978.

E.g. For 1979 the percentage price relative to that in 1978 is 125%.

For 1976 the percentage price relative to that in 1978 is 66.6%.

When calculating percentage price relatives, it is usual to miss out the word percentage and the percentage signs.

So we have for 1979 the price relative to 1978 is 125.

for 1976 the price relative to 1978 is 66.6.

The price relative for the base year is taken as 100.

In our example the price relative for the base year 1978 is 100.

Price relatives are examples of *index* numbers.

Other examples of index numbers include wage relatives, share relatives, production relatives, etc.

Combining Index Numbers into a Composite Index

The index numbers of a group of items can be combined into a composite index number for those items.

E.g. the price relatives for a group of commodities can be combined into a composite price relative for those commodities.

A composite index number gives the *overall* index number for a group of items.

E.g. a composite price relative gives the overall price relative for a group of commodities.

A composite index number is calculated as a *weighted* average of the separate index numbers of the various items.

Example: The index numbers (price relatives) for commodities B, C, D, E and F in 1979 were 105, 103, 110, 115 and 108, compared with a base year of 1977. The commodities B, C, D, E and F have weightings of 5, 6, 2, 3 and 4. Find the composite index of the commodities in 1979.

	x Index	w Weight	wx Index × Weight
Commodity B	105	5	525
Commodity C	103	6	618
Commodity D	110	2	220
Commodity E	115	3	345
Commodity F	108	4	432

$$\Sigma w = 20 \qquad \Sigma wx = 2140$$

$$\Sigma wx = 2140$$
$$\Sigma w = 20$$
$$\text{Composite Index} = \frac{\Sigma wx}{\Sigma w}$$
$$= \frac{2140}{20}$$
$$= 107$$

The composite index of 107 shows that *overall* the price of the commodities rose by 7% between 1977 and 1979.

The two composite index numbers with which ordinary people are acquainted are the Index of Retail Prices and the Financial Times Share Index.

The Index of Retail Prices is calculated and published by the government each month.

The Financial Times Share Index, which is also known as the F.T. Index, is calculated and published by that newspaper each day. The F.T. Index is also broadcast on radio and television each day.

Index of Retail Prices

The Index of Retail Prices is calculated as a weighted mean of the price relatives of the various items of expenditure which are considered as necessary to each household.

The items of expenditure considered necessary are broadly divided into the following categories.

food	fuel and light	miscellaneous
alcohol	durable household goods	services, e.g. meals outside
tobacco	clothing	the home.
housing	transport	

The sum of the weightings given to each of the categories is 1,000.

The *total* of the weights is 1,000.

The weightings given to each category are continually reviewed and are based on the actual amount of money spent on each category throughout the country.

The weights used during any year are based on the actual amounts spent on each category during the preceding three years.

E.g. if for every £1,000 spent throughout the country, surveys show that £300 was spent on food, £60 on alcohol, £65 on tobacco, £120 on housing and so on, the weightings given are food 300, alcohol 60, tobacco 65 and housing 120.

The following table shows the calculation of the Index of Retail Prices for June of year X, taking January in year Y as base. A guess of 125 has been taken to reduce the calculations.

Category	x Index	w Weight	g	d $x-g$	wd $w(x-g)$
Food	125	280	125	0	0
Alcohol	115	60	125	−10	−600
Tobacco	120	70	125	−5	−350
Housing	140	140	125	15	2100
Fuel	145	70	125	20	1400
Goods	120	65	125	−5	−325
Clothing	110	90	125	−15	−1350
Transport	130	100	125	5	500
Misc.	120	75	125	−5	−375
Services	125	50	125	0	0

Guess = 125
Σwd = 1000
Σw = 1000

$\Sigma w = 1000$

$4000 - 3000$
$\Sigma wd = 1000$

Index of R.P. $= \dfrac{\Sigma wd}{\Sigma w} + g$

$= \dfrac{1000}{1000} + 125$

$= 1 + 125$

$= 126$

Miscellaneous Exercises 11

1. A pupil scores the following marks in a test:

English 70,   mathematics 60,   history 52.

Calculate the pupil's weighted average mark, if the weightings used are English 4, mathematics 3, history 1.   WM78

2. A pupil scores the following marks in a test:

English 60,   mathematics 40,   geography 46

Calculate the pupil's weighted average mark, if the weightings used are English 3,   mathematics 2,   geography 1   WM76

3. In the table below, the marks of three candidates, A, B and C, are given for each of three tests they took when applying for a job. The weights assigned to each test are also given.

	Test I	Test II	Test III
Marks of candidate A	62	80	44
Marks of candidate B	60	72	54
Marks of candidate C	80	62	44
Weight	5	4	3

(a) Calculate:
    (i) the mean mark for each candidate
    (ii) the weighted mean mark for each candidate.
(b) Explain briefly why a weight was assigned to each test.
(c) On the basis of your results in part (a), state, with a reason:
    (i) which applicant was most likely to have been given the job
    (ii) whether you think this was a fair or an unfair decision.   M78

4. In a scholarship examination the examiners decided to give different weights to the marks gained for each of five papers. In the table below is given a list of marks obtained by two candidates, P and Q, and the weight assigned to each paper.

	Paper				
	I	*II*	*III*	*IV*	*V*
Marks of candidate P	57	71	55	69	98
Marks of candidate Q	75	77	79	67	52
Weight	6	5	4	3	2

Considering the five marks awarded for each candidate, work out:

(a) the ordinary mean mark for each of candidates P and Q *before* the weights are applied,

(b) the *weighted* mean mark of each of the candidates P and Q.

Find the difference, for each of the candidates, P and Q, between his ordinary mean mark and his weighted mean mark. Comment on any surprising feature of these differences.   M76

5. The index number for the price of a commodity in 1974, taking the 1972 price as base, is 250. What is the 1972 index number taking the 1974 price as base?   WM75

6. The index number for the price of a commodity in 1976 taking the 1975 price as base is 125. What is the 1975 index number taking the 1976 price as base?   WM76

7. In the Index of Retail Prices, the item food has a weight of 314, whilst the item alcoholic drink has a weight of 63. State *briefly* why the first weight is so much larger than the second.   EA78

8. In January 1974, the Index for Fuel and Light was 100. A firm's bill for fuel and light for the first quarter of 1974 was £275. In the first quarter of 1977, using approximately the same amount of fuel and light, the firm received a bill for £550. Estimate the Index for Fuel and Light in January 1977.   EA78

9. In July 1977 the Food Index was 192, based on January 1974 = 100. Approximately how much money would be needed in July 1977 to pay for a parcel of groceries which, in January 1974, would have cost £25?
EA78

10. The figures below show how much Christmas dinner would have cost a family in 1935 and in 1975.

    (*a*) Complete the following menus:

1935	£	p	1975	£	p
11 lb turkey at 6p per lb			11 lb turkey at 55p per lb		
1½ lb pork sausage at 3p per lb			1½ lb pork sausage at 38p per lb		
Sausagemeat and herb stuffing		6	Sausagemeat and herb stuffing		38
2 lb potatoes		1	2 lb potatoes		16
2 lb Brussels sprouts		2	2 lb Brussels sprouts		24
Brandy sauce		1½	Brandy sauce		25
2 lb Christmas pudding		10	2 lb Christmas pudding		68
Total			Total		

(*b*) Calculate the difference between the total costs of the two menus.
(*c*) The 1935 and 1975 prices for the Christmas puddings were as shown. Calculate the increase as a percentage of the 1935 price. Y77

11. (*a*) Construct a composite index number, as a weighted mean, from the following data:

Commodity	A	B	C	D	E
Index Number	124	153	98	133	107
Weight	9	5	4	1	6

(*b*) The mean of a set of 5 numbers is 6, and their standard deviation is 3. Two more numbers 2 and 10 are added. Calculate the standard deviation of the new set.   WM76

12. A composite index number is to be constructed from the following index numbers, weighted as shown. Calculate the composite index number as a weighted arithmetic mean.

Index number	170	145	130	125	110	105
Weight	3	23	45	16	8	5

M77

13. (*a*)

	Price Relative	Weight
Food	125	4
Rent, rates	119	2
Clothing	135	1
Fuel	186	1
Household goods	122	0.5
Miscellaneous	116	1.5

Use the data given above to work out a cost of living index.

(*b*) A set of 8 numbers has a mean value of 20 and a standard deviation of 3. Two more numbers 8 and 32 are added to the set, leaving the mean value unchanged.

Find the standard deviation of the new set of 10 numbers. WM78

12 PROBABILITY 2

Combining Probabilities

Mutually exclusive events

Two or more events are *mutually exclusive* when the occurrence of one of the events excludes the occurrence of any of the other events, i.e. if one of the events happens then the others cannot.

E.g. the outcomes when a die is tossed once are 1, 2, 3, 4, 5, 6.

These outcomes are mutually exclusive since if the die shows any one of them it cannot show any of the others.

If the die shows a 2, it is not possible for it to show 1, 3, 4, 5 or 6.

Independent events

Two or more events are *independent* if none of the events can affect any of the other events in any way, i.e. they are independent if the knowledge that one has occurred has no effect on the probabilities of the others occurring.

E.g. when a coin and a die are tossed together the result obtained on the coin has no effect whatsoever on the result obtained on the die and vice-versa.

The probability that a head appears on the coin is $\frac{1}{2}$ and the result shown on the die cannot affect this probability.

Similarly the probability that a 3 is shown on the die is 1/6 and the result shown on the coin cannot affect this probability.

Dependent events

Two or more events are *dependent* if the knowledge that one of the events has occurred alters the probabilities of the other events occurring.

E.g. the probability of drawing an ace from a full pack of cards is 4/52 = 1/13.

When 2 cards are drawn from a pack of cards one after the other, the probability that the first card is an ace is 1/13.

Assuming that the first card is an ace and if the card is *not* replaced in the pack, the probability that the second card is an ace is *not* 1/13.

After the first card has been drawn there are only 3 aces left in a pack of 51 cards so the probability that the second card is an ace is 3/51. The knowledge that the first draw was an ace has altered the probability that the second card drawn will be an ace.

It is often necessary to calculate the probabilities of multiple events which occur as a result of the combination of a number of happenings.
It is possible to calculate such probabilities by constructing composite sample spaces for the happenings as described on pages 162–4.
The construction of a composite sample space can be very slow and tedious, especially if the number of happenings is large.
The following approach eliminates the need for complex composite sample spaces.

A multiple event consists of two or more separate events, e.g. the probability of obtaining a 6, a head and a pin pointing up when a die, coin and drawing pin are tossed together might be required.
The multiple event required is 6, head and point up. This event is made up of three separate events, 6 on the die, head on the coin, and point up on the drawing pin.
The probabilities of the separate events which make up a multiple event can be *combined* together to give the probability of the multiple event.

If the events which make up a multiple event are entirely *mutually exclusive* to each other, the probabilities of the separate events can be simply *added* together.

If the events which make up a multiple event are entirely *independent* of each other, the probabilities of the separate events can be simply *multiplied* by each other.

If the events which make up a multiple event are entirely *dependent* on each other, the probabilites of the separate events can be simply *multiplied* by each other.

If the separate events which make up a multiple event are a mixture of mutually exclusive, independent and/or dependent events, it may be necessary to use *both* addition and multiplication to combine the separate probabilities of the separate events.
To find the probability of a multiple event:

    (*a*) decide on the types of event present.
    (*b*) calculate the probabilities of each of the separate events.
    (*c*) combine the separate probabilities obtained in (*b*) by the method(s) dictated by the answers given by (*a*).

The decision regarding the types of event present may not be easy to make. Some deep thought may be required, backed up by much practice, but the situation is frequently eased by the wording of a question.

Find the probability of:

Either .... or .... indicates that the events are mutually exclusive, hence *add*.

Find the probability of:

Both .... and .... indicates that the events are independent or dependent, hence *multiply*.

Example: Find the probability that when a die, coin and a drawing pin are thrown together, the result is a 6, a head and point up.

(*a*) The events are independent $\therefore$ multiply.

(*b*) $P(6) = \frac{1}{6}$ $P(H) = \frac{1}{2}$ $P(\perp) = \frac{2}{3}$ (approx.)

(*c*) $P(6 \text{ and } H \text{ and } \perp) = P(6) \times P(H) \times P(\perp)$
$$= \frac{1}{6} \times \frac{1}{2} \times \frac{2}{3} = \frac{2}{36} = \frac{1}{18}.$$

Example: Find the probability that when a die is tossed once the result will be either a 4 or 5.

(*a*) Events are mutually exclusive $\therefore$ add.

(*b*) $P(4) = \frac{1}{6}$ $P(5) = \frac{1}{6}$

(*c*) $P(4 \text{ or } 5) = P(4) + P(5)$
$$= \frac{1}{6} + \frac{1}{6} = \frac{2}{6} = \frac{1}{3}.$$

Example: Find the probability of drawing an ace from a pack of cards and throwing a head at the first attempt.

(*a*) The events are independent $\therefore$ multiply.

(*b*) $P(\text{ace}) = \frac{4}{52} = \frac{1}{13}$ $P(\text{head}) = \frac{1}{2}$

(*c*) $P(\text{ace and head}) = P(\text{ace}) \times P(\text{head})$
$$= \frac{1}{13} \times \frac{1}{2} = \frac{1}{26}.$$

Example: Find the probability of drawing 3 aces in succession from a pack of cards if the cards are *not* replaced after each draw.

(*a*) The events are dependent $\therefore$ multiply.

(*b*) $P(\text{ace } 1) = \frac{4}{52} = \frac{1}{13}$
$P(\text{ace } 2) = \frac{3}{51} = \frac{1}{17}$
$P(\text{ace } 3) = \frac{2}{50} = \frac{1}{25}.$

(*c*) $P(\text{ace 1 and ace 2 and ace 3})$
$$= P(\text{ace } 1) \times P(\text{ace } 2) \times P(\text{ace } 3).$$
$$= \frac{1}{13} \times \frac{1}{17} \times \frac{1}{25} = \frac{1}{5525}.$$

Example: Find the probability of drawing either a heart or a spade from a pack of cards at the first draw.

(*a*) The events are mutually exclusive ∴ add.
(*b*) P(heart) = $\frac{13}{52} = \frac{1}{4}$ P(spade) = $\frac{13}{52} = \frac{1}{4}$
(*c*) P(heart or spade) = P(heart) + P(spade)
$$= \tfrac{1}{4} + \tfrac{1}{4} = \tfrac{2}{4} = \tfrac{1}{2}.$$

Example: Find the probability of forecasting correctly the results of 6 given football matches on a particular Saturday afternoon.

(*a*) The events are independent ∴ multiply.
(*b*) Each match can finish home win, away win or draw.
    P(correct forecast for any match) = P(C) = $\frac{1}{3}$.
(*c*) P(correct forecasts for all six matches)
$$= P(C) \times P(C) \times P(C) \times P(C) \times P(C) \times P(C).$$
$$= \tfrac{1}{3} \times \tfrac{1}{3} \times \tfrac{1}{3} \times \tfrac{1}{3} \times \tfrac{1}{3} \times \tfrac{1}{3} = \tfrac{1}{729}.$$

Example: Find the probability of drawing a king or an ace or a queen from a pack of cards at the first attempt.

(*a*) The events are mutually exclusive ∴ add.
(*b*) P(king) = $\frac{4}{52} = \frac{1}{13}$ P(queen) = $\frac{4}{52} = \frac{1}{13}$ P(ace) = $\frac{4}{52} = \frac{1}{13}$
(*c*) P(king or queen or ace) = P(king) + P(queen) + P(ace).
$$= \tfrac{1}{13} + \tfrac{1}{13} + \tfrac{1}{13} = \tfrac{3}{13}.$$

A multiple event may consist of a *mixture* of mutually exclusive events, independent events and dependent events.
A multiple event of this type necessitates the addition and multiplication of probabilities in the same calculation.

Example: Find the probability that when a card is drawn from a pack of cards and a die is tossed the final result is either an ace or a king from the cards together with a six or a three on the die.

(*a*) The event required is a mixture of mutually exclusive and independent events.
(*b*) P(ace) = $\frac{4}{52} = \frac{1}{13}$ P(king) = $\frac{4}{52} = \frac{1}{13}$
    P(6) = $\frac{1}{6}$ P(3) = $\frac{1}{6}$
(*c*) P([ace or king] and [6 or 3])
$$= P(\text{ace or king}) \times P(6 \text{ or } 3).$$
$$= [P(\text{ace}) + P(\text{king})] \times [P(6) + P(3)]$$
$$= [\tfrac{1}{13} + \tfrac{1}{13}] \times [\tfrac{1}{6} + \tfrac{1}{6}]$$
$$= \tfrac{2}{13} \times \tfrac{2}{6}$$
$$= \tfrac{4}{78} = \tfrac{2}{39}$$

Example: Find the probability that when a card is drawn from a pack of cards and a coin is tossed, the final result is either an ace and a head or a king and a tail.

(a) The required event is a mixture of mutually exclusive and independent events.

(b) $P(\text{ace}) = \frac{1}{13}$ $P(\text{head}) = \frac{1}{2}$ $P(\text{king}) = \frac{1}{13}$ $P(\text{tail}) = \frac{1}{2}$.

(c) $P([\text{ace and head}] \text{ or } [\text{king and tail}])$
$$= P(\text{ace and head}) + P(\text{king and tail})$$
$$= [P(\text{ace}) \times P(\text{head})] + [P(\text{king}) \times P(\text{tail})]$$
$$= [\tfrac{1}{13} \times \tfrac{1}{2}] + [\tfrac{1}{13} \times \tfrac{1}{2}]$$
$$= \tfrac{1}{26} + \tfrac{1}{26} = \tfrac{2}{26} = \tfrac{1}{13}.$$

Tree Diagrams

Tree diagrams are very useful when calculating the probabilities of multiple events.

Each happening can be represented by a cluster of branches.

E.g. the toss of a die is represented by its own cluster of branches.

It is not necessary for each outcome of a happening to have a separate branch. Only those events which are part of the required multiple event need their own branch in the cluster of the happening.

All of the other outcomes of a happening can be grouped together and represented by one branch of the cluster.

Grouping some of the outcomes together gives a condensed sample space for a happening.

If a happening X can have the outcomes A, B, C, D, and E and if only event D is part of a required multiple event then happening X can be represented by the cluster:

D   condensed sample space since outcomes
Not D   A, B, C, E are grouped together as *not* D.

The branch *not* D is representing all of the outcomes A, B, C and E. *Not* D can be shortened to $\overline{D}$,

this gives the cluster

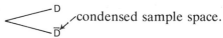

D   condensed sample space.
$\overline{D}$

E.g. the outcomes when a die is tossed are 1, 2, 3, 4, 5, 6.

If a score of 4 is part of a required multiple event, then the toss of a die can be represented by the cluster

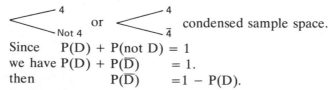

4
Not 4   or   4
$\overline{4}$ condensed sample space.

Since $P(D) + P(\text{not } D) = 1$
we have $P(D) + P(\overline{D})$ $= 1$.
then $P(\overline{D})$ $= 1 - P(D)$.

Each branch of a tree diagram represents an event and each one of those events has a probability,
E.g. for the cluster below representing the toss of a coin

$$\begin{array}{c} \diagup\; 4 \\ \diagdown\; \bar{4} \end{array}$$

we have $P(4) = \frac{1}{6}$ and $P(\bar{4}) = 1 - P(4) = 1 - 1/6 = 5/6$.
The probabilities of each event can be written next to the branch representing that event,
e.g.

$$\begin{array}{c} {}_{1|6}\diagup\; 4 \\ {}_{5/6}\diagdown\; \bar{4} \end{array}$$

Similarly for a draw from a pack of cards
$$P(\text{ace}) = \tfrac{1}{13} \text{ and } P(\text{not ace}) = P(\overline{\text{ace}}) = 1 - P(\text{ace})$$
$$= 1 - 1/13 = 12/13.$$

giving the cluster

$$\begin{array}{c} {}_{1|13}\diagup\; \text{Ace} \\ {}_{12/13}\diagdown\; \overline{\text{Ace}} \end{array}$$

A tree diagram for a multiple event is built up in stages as described on page 160, but using clusters showing condensed sample spaces and complete with the probabilities of each branch.
The tree diagram produces a condensed composite sample space.
Each pathway through the tree diagram gives an event in the condensed composite sample space.
The probability of each event in the condensed composite sample space can be found by *multiplying* together all of the probabilities of the branches leading to that event.
E.g. For the path shown below where $P(B) = 1/3$ and $P(A) = 2/5$.

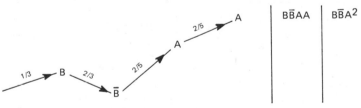

The final event is $B\bar{B}A^2$.
$$P(B\bar{B}A^2) = \tfrac{1}{3} \times \tfrac{2}{3} \times \tfrac{2}{5} \times \tfrac{2}{5} = \tfrac{8}{225}$$
If a required event occurs more than once in the composite sample space, the probabilities of each of the times it occurs are *added* together to give the total probability of the required event.

Example: Find the probability that when a card is drawn from a pack of cards and a die is tossed, the final result is either an ace or king from the cards, together with a six or a three on the die.

Cluster for cards is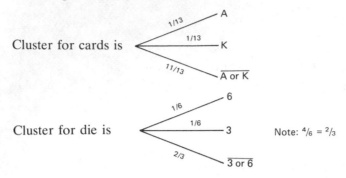

Cluster for die is

Note: $^4/_6 = ^2/_3$

Combining into a tree diagram gives

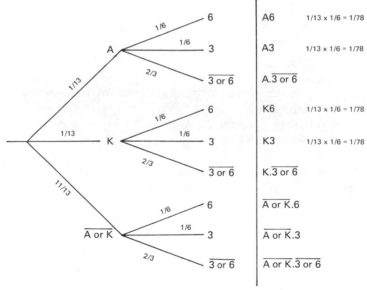

A6	1/13 x 1/6 = 1/78
A3	1/13 x 1/6 = 1/78
A.$\overline{3 \text{ or } 6}$	
K6	1/13 x 1/6 = 1/78
K3	1/13 x 1/6 = 1/78
K.$\overline{3 \text{ or } 6}$	
$\overline{\text{A or K}}$.6	
$\overline{\text{A or K}}$.3	
$\overline{\text{A or K}}$.$\overline{3 \text{ or } 6}$	

The events A6, A3, K6, K3 all satisfy the required event, so their probabilities are added together giving
P([ace or king] and [6 or 3]) = 1/78 + 1/78 + 1/78 + 1/78
= 4/78 = 2/39

Example: If there are 10 balls in a bag and 3 of them are black and 7 are white, what is the probability that if 2 balls are drawn out of the bag they are:

(a) both white?
(b) one of each colour?
(c) both black?

The cluster for the first ball is

$$3/10 \quad B$$
$$7/10 \quad W$$

There are two clusters for the second ball depending on whether a black ball was chosen first or a white ball was chosen first (*dependent* events).

Black ball first gives

$$2/9 \quad B$$
$$7/9 \quad W$$

for second ball.

White ball first gives

$$3/9 \quad B$$
$$6/9 \quad W$$

or

$$1/3 \quad B$$
$$2/3 \quad W$$

for second ball.

Combining these clusters into a tree diagram gives

3/10 B	2/9 B	BB	B^2	3/10 x 2/9 = 6/90 = 1/15
	7/9 W	BW	BW	3/10 x 7/9 = 21/90 = 7/30
7/10 W	1/3 B	WB	BW	7/10 x 1/3 = 7/30 = 7/30
	2/3 W	WW	W^2	7/10 x 2/3 = 14/30 = 7/15

(*a*) $P(W^2) = 7/15$.

(*b*) There are two ways of obtaining one of each colour, i.e. BW.
    $P(\text{one of each colour}) = 7/30 + 7/30 = 14/30 = 7/15$.

(*c*) $P(B^2) = 1/15$.

Miscellaneous Exercises 12

1. A bag contains 3 red and 5 green balls. A ball is selected at random from the bag and thrown away, then a second ball is selected.

The tree diagram below shows the probabilities of selecting red or green balls from the bag.

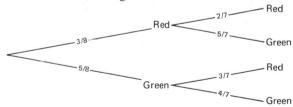

The probability of selecting 1 red and 1 green ball from the bag is

(*a*) $(\frac{3}{8} \times \frac{5}{7})$

(*b*) $(\frac{3}{8} \times \frac{5}{7}) + (\frac{5}{8} \times \frac{3}{7})$

(*c*) $(\frac{3}{8} + \frac{5}{7})$

(*d*) $(\frac{3}{8} + \frac{5}{7}) \times (\frac{5}{8} + \frac{3}{7})$

Y78

2. Two bags contain coloured marbles. Bag A contains 3 red marbles and 2 yellow marbles, and bag B contains 1 red marble and 4 yellow ones.

One marble is chosen at random from bag A, followed by one at random from bag B.

A tree diagram has been started below for the outcomes and probabilities for these two draws.

1st draw	2nd draw	outcome	probability
	R	(R, R)	$\frac{3}{5} \times \frac{1}{5} = \frac{3}{25}$
3/5 — R			
Y			
	Y	(Y, Y)	

(*a*) Copy and complete the table.

(*b*) Find the probability of having at least one red marble in the two marbles selected.

(*c*) Calculate the probability of having a final outcome of (Y, Y, Y) if another draw is made from bag B without replacing either of the marbles drawn so far.   M78

3. Two bags contain coloured marbles. Bag A contains 1 red and 2 green marbles. Bag B contains 3 red and 5 yellow marbles. One marble is chosen at random from bag A followed by one marble from bag B. An incomplete tree diagram is given below for the outcomes and probabilities for the two draws.

1st draw	2nd draw	outcome	probability
	3/8 — R	RR	$\frac{1}{3} \times \frac{3}{8} = \frac{1}{8}$
1/3 — R			
	Y		
2/3 — G	R		
	5/8 — Y	GY	

(*a*) Copy and complete the tree diagram by inserting the missing outcomes and probabilities.

(*b*) Find the probability of having at least one red marble in the outcome of the two draws.

(*c*) Assuming that the marbles are not replaced after each draw calculate

the probability of having a final outcome (R, R, R) following another
draw from bag B.   M77

4. (*a*) In a class of 30 pupils, 6 wear glasses. If two pupils are selected at
    random, what is the probability that both wear glasses if they are
    selected:
    (i) with replacement?
    (ii) without replacement?

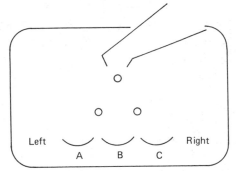

(*b*) In the game illustrated in the diagram, a ball falls and is deflected by
    two rows of pins to arrive in one of three cups, A, B or C. If the
    probability that the ball is deflected to the left of each pin is $\frac{2}{3}$,
    (i) what is the probability that it is deflected to the right of each pin?
    (ii) Illustrate the probabilities of route by using a tree diagram.
    (iii) If 1,000 balls fall through the pins, how many would you expect
        to find in each cup?   NI77

5. Three ten pence coins are tossed together. One possible way that they can
   land is Head, Tail, Head, indicated on the table below by H, T, H.

   Complete the table below to show the other 7 different ways in which the
   coins could land.

Coin 1	Coin 2	Coin 3
H	T	H

(*a*) From your table, write down the probability of:
  (i) the three coins all falling Heads,
  (ii) two coins falling Heads and one coin Tail.
(*b*) The third coin is now changed for a six-sided die and the two remaining coins and the die are tossed together.
Calculate the probability that the throw will give:
  (i) tail, tail, six,
  (ii) one head, one tail and either a 1 or a 2.   Y77

6. The pattern of possibilities in forecasting the results of a number of football matches is illustrated below, where H stands for a 'home' win, A stands for an 'away' win and D for a 'draw'.

Number of matches forecast	The ways that the results can be forecast			Total number of possibilities
1	H,	A,	D	3
2	HH,	HA,	HD	9
	AH,	AA,	AD	
	DH,	DA,	DD	

(*a*) Using the above as a guide, write in your answer book a table like the one above to show the pattern of possibilities obtained when the forecasts are made for *three* matches (beginning with HHH and ending with DDD).
(*b*) Use the pattern of possibilities given, and the one you have just found, to give answers to the following questions, commenting, where necessary, on any assumptions that you make.
  (i) When the result of two matches is forecast, what is the chance that a forecast will:
    (1) consist of one home win and one away win
    (2) include at least one draw
    (3) *not* include a home win?
  (ii) When the result of three matches is forecast, what is the chance that the forecast will:
    (1) be three home wins
    (2) consist of two home wins and a draw
    (3) consist of one draw and two other results, neither of which is a draw?   M77

7. When a single coin is tossed or when two similar coins are tossed together, the pattern of possibilities is illustrated below, where H stands for 'heads' and T stands for 'tails'.

Number of coins	The ways the coin or coins can come down	Total number of possibilities
1	H; T	2
2	HH; HT; TH; TT	4

Using the above table as a guide, write in your answer book a table like the one above to show the possibilities obtained first when three similar coins are tossed together and then when four similar coins are tossed together.

Use the patterns of possibilities given, and the ones that you have just found, to give answers to the following questions.

(*a*) When two similar coins are tossed together, what is the chance that the coins will fall with:
  (i) no heads upwards,
  (ii) *at least* one tail upwards?

(*b*) When three similar coins are tossed together, what is the chance that the coins will fall with:
  (i) two heads and one tail upwards,
  (ii) *not less* than one head upwards?

(*c*) When four similar coins are tossed together, what is the chance (*in its simplest form*) that the coins will fall with:
  (i) three heads and one tail upwards,
  (ii) two heads and two tails upwards?   M76

8. (*a*) Copy and then complete rows 4 and 5 of this pattern based on 'Pascals' triangle.

```
Row No. 1              1    1
Row No. 2          1    2    1
Row No. 3      1    3    3    1
Row No. 4
Row No. 5
```

(*b*) If 3 coins are tossed at the same time, what are the probabilities that:
  (i) they will all land with 3 heads showing?
  (ii) they will all land showing the same side?
  (iii) they will land with 2 heads and 1 tail showing?

(*c*) (i) What is the probability of throwing a 3 or a 4 with one roll of a die?
  (ii) What is the probability of throwing not more than 5 with one throw of a die?

(*d*) A bag contains 5 black and 3 red marbles. The marbles are to be taken out two at a time. What is the probability of drawing out a red and black marble, together, the first time any marbles are taken out?

(*e*) If a set of results are 'Normally' distributed,
  (i) *sketch* the type of graph you would expect to obtain from these results,
  (ii) indicate on your sketch an ordinate representing the Mean of these results.
(*f*) Write down the symbol used for Standard Deviation.   NW74

13 SAMPLING TECHNIQUES

Sampling Methods

Obtaining random samples tends to be very expensive.
There are cheaper methods of obtaining samples which although *not* giving *truly* random samples do give sufficiently random samples for most surveys.
The four most commonly used methods of obtaining samples are:

(*a*) systematic sampling
(*b*) stratified sampling
(*c*) multi-stage sampling
(*d*) quota sampling.

A *systematic sample* is one chosen by (*a*) taking a list of all of the population (*b*) picking the first member of the sample at random and (*c*) picking the rest of the sample by taking every 100th (say) member in the list both before and after this first member.

A *stratified sample* is one chosen by picking at random 10% (say) of each identifiable section or strata of the population.

A *multi-stage sample* is one chosen one step at a time.
Starting with large groupings, smaller groupings are chosen at random from these large groupings. Even smaller groupings are then chosen at random and so on until the individual members of the sample are identifiable.
E.g. if the population is all the voters in Great Britain, a sample could be chosen by (*a*) first picking out a given number of counties, (*b*) then picking out a given number of towns in each of the chosen counties, (*c*) then picking out a number of voting wards in each of the chosen towns, (*d*) then picking out a given number of voters from the electoral registers of each of the chosen wards.
Note: At each stage the selections must be random.

A *quota sample* is one chosen by (*a*) deciding how many of each type or category of member should be in the sample, e.g. 20 married men, 20 married women, 15 single men, 10 housewives, 15 single women, 5 children between 8 and 12 years of age etc. (*b*) then going out and finding the correct number of each type of member required for the sample.
Note: A person could fit more than one category, e.g. married woman and housewife. She would be counted twice, once in the married women category and once in the housewives category.

Example: Consider a school of 1,200 pupils with 200 pupils in each year from the 1st year to the 6th year. The school has 8 classes of 25 pupils in each year. The 1st year has 80 boys and 120 girls, the 2nd year has 100 boys and 100 girls, the 3rd year has 110 boys and 90 girls, the 4th, 5th, and 6th years each have 100 boys and 100 girls.

To obtain a 10% systematic sample for the above school:

   (*a*) obtain a full set of class lists.
   (*b*) arrange class lists in order, i.e. 1A, 1B, 1C . . . up to 6G, 6H.
   (*c*) number all of the pupils on the first class list with the numbers 1–25.
   (*d*) put the numbers 1–25 in a hat.
   (*e*) pick out one number.
   (*f*) find the pupil on the list from (*c*) with that number.
   (*g*) take every 10th pupil in the lists starting from the pupil found in (*f*).

To obtain a 10% stratified sample for the school:

   (*a*) obtain a full set of class lists.
   (*b*) number the pupils in the first year with the numbers 1–200.
   (*c*) put the numbers 1–200 in a hat.
   (*d*) pick out 20 numbers.
   (*e*) find the 20 pupils with these numbers on the first year class lists.
   (*f*) repeat steps (*b*) to (*e*) for the 2nd, 3rd, 4th, 5th, 6th years.

Note: $10\% = \frac{1}{10}$th. $\frac{1}{10}$th of 200 = 20.

One way, although somewhat unrealistic, to obtain a 10% multi-stage sample for the above school is to:

   (*a*) obtain a full set of class lists.
   (*b*) number the lists with the numbers 1–48 in black (say).
   (*c*) put the numbers 1–48 in a hat and pick out 24 numbers.
   (*d*) take the lists with these numbers and renumber them 1–24 in red (say).
   (*e*) put numbers 1–24 in a hat and pick out 12 numbers.
   (*f*) take the lists with these *red* numbers.
   (*g*) number the pupils on each of these lists with the numbers 1–25 in green (say).
   (*h*) put the numbers 1–25 in a hat and pick out 10 numbers.
   (*i*) find the pupils on one of the lists from (*g*) with these numbers.
   (*j*) repeat steps (*h*) and (*i*) until all of the lists in (*g*) have been completed.

To obtain a 10% quota sample for the above school:

   (*a*) obtain a full set of class lists. *Note:* 10% of 1,200 = 120.
   (*b*) work out the number of each type of pupil required in the sample.

The sample will require 120 pupils, i.e. 20 pupils per year.

1st year has 80 boys : 120 girls or 4 : 6 or 8 : 12
2nd year has 100 boys : 100 girls or 1 : 1 or 10 : 10
3rd year has 110 boys : 90 girls or 11 : 9 or 11 : 9
4th year has 100 boys : 100 girls or 1 : 1 or 10 : 10
5th year has 100 boys : 100 girls or 1 : 1 or 10 : 10
6th year has 100 boys : 100 girls or 1 : 1 or 10 : 10

(c) at break go into the playground and find:

   8 1st year boys   12 1st year girls
  10 2nd year boys   10 2nd year girls
  11 3rd year boys   9 3rd year girls
  10 4th year boys   10 4th year girls
  10 5th year boys   10 5th year girls
  10 6th year boys   10 6th year girls.

(d) find out the names of the pupils found in (c) and mark them on the class lists.

Methods of Collecting Data

Data is collected in many different ways. The main ways are:

(a) census
(b) postal questionnaire
(c) enumerator questionnaire
(d) by observation
(e) by experiment
(f) from published statistics.

Census

Every 10 years or so the Government conducts a survey of the whole population of the country. Every household is given a census form which it must by law fill in and return. The form contains a series of questions, the answers to which are of interest to the government, e.g. (a) how many people are there in each household? (b) how old are they? (c) how much do they earn? etc.

Since a census form must be answered and returned by every household, conducting a census is a very expensive undertaking and it takes a long time to complete.

Questionnaire (Postal)

Many firms and institutions need detailed information about public reaction to their products or ideas so that they can plan their future developments, e.g. (*a*) what features do the public like? (*b*) what features do the public dislike? (*c*) what alterations would the public like to see? etc.

Mounting a census to obtain the information would take too long and be much too expensive. The solution is to mount a kind of mini-census.

A sample of the population is chosen and a questionnaire (list of questions) is sent to all of the people in the sample. The cost is kept low and the survey can be completed in a relatively short space of time. The big disadvantage of questionnaires is that since people do not by law have to fill them in, they are frequently not returned by a large number of people in the sample. A 20% return is considered to be very good. This leads to bias, as only interested people or dissatisfied people bother to reply.

Questionnaire (Enumerator)

Enumerator questionnaires serve exactly the same purpose as postal questionnaires. The difference between the two types of questionnaire is that the questionnaires are not sent out to a sample but instead interviewers (enumerators) go to see the members of the chosen sample. The interviewers ask the necessary questions of the people in the sample and fill in the answers themselves.

This method ensures that all of the questionnaires are all filled in, unless of course a member of the sample refuses to co-operate. The cost can be kept low and the survey can be completed in a relatively short space of time. Enumerator questionnaires are frequently used in public opinion polls when the results are required almost immediately, e.g. who is likely to win the next election? For public opinion polls quota sampling is used, i.e. people fitting the right description of those who should be interviewed are stopped in the street and questioned.

By Observation

Some data cannot be obtained by questioning, e.g. traffic flows in a town. To collect data of this kind, the people who require the data, e.g. town planners, must go and observe exactly what does happen at a set of traffic lights (say). Every type of event and how many times it happens must be noted carefully. E.g. at a set of traffic lights (*a*) how long do the lights stay on red? (*b*) how long do the lights stay on green? (*c*) how many cars turn left? (*d*) how many go straight on? (*e*) how many turn right? (*f*) how often do traffic jams occur?

By Experiment

Some data can only be obtained by doing experiments, e.g. the length of time a type of light bulb will last, the effect of a fertilizer on the yield of a crop, the variability in the weights of sugar bags.

To find the lifetime of a type of light bulb, a sample of the bulbs would be tested to destruction, i.e. until they failed.

From Published Statistics

Many statistics are collected either by law or as a matter of routine by government agencies. These statistics are published in a monthly gazette. Any information which is of value to interested bodies can be easily extracted from this gazette.

Note: This type of data is called *secondary data* as it has not been collected by the person extracting and using it.

What makes a Good Questionnaire?

When making up a questionnaire, the following points should be kept in mind.

A questionnaire should not be too long.

A questionnaire should contain all of the questions required to cover the purpose of the survey.

Each question should be phrased in very simple language so that everyone can understand the question.

Each question must be phrased so that it can have only one possible meaning.

Each question should be capable of being answered by a very short precise answer, e.g. yes or no, a number, or by ticking a box.

Vague descriptive words should be avoided, e.g. when does small become large?

Questions should not require the person answering them to do any calculations, as he may make a mistake in the calculations.

Leading questions, i.e. those which might suggest an answer, should not be asked.

Questions which double check on the answers to other questions should be included.

Questions which could cause embarrassment should be avoided.

A series of short questions can often be used to eliminate:

    (*a*) vague words
    (*b*) the need for calculations.

A Statistical Survey

Any statistical investigation should follow a general plan. Such a plan is listed below.

(1) Decide on the aims of the survey.
(2) Decide what data is essential and what methods are necessary for collecting it.
(3) Select a sample when necessary.
(4) Work out the questions for a questionnaire when one is to be used.
(5) Collect the data.
(6) Process the data.
(7) Present the findings of the survey by suitable methods.

Miscellaneous Exercises 13

1. Frequently in statistics we need to select a sample. Give *one* reason why this may be necessary.   WM77

2. Define a random sample.   EA78

3. Imagine you are in a large mixed comprehensive school, and are asked to find out which television programmes are most popular with the pupils. You decide to ask a limited number of pupils to fill in questionnaires, using a method of stratified sampling to select the pupils. State *very briefly* how you would obtain your samples.   EA78

4. In a certain school a random sample of 20 pupils is required.
Give *one* reason why going to the gate in the morning and selecting the first 20 pupils to arrive is likely to give biased results.   WM76

5. Give *one* reason why a survey conducted by questioning customers outside a supermarket is likely to give biased results.   WM78

6. A survey on reading habits, investigating the number and type of books read by individuals, is conducted by questioning customers at a public library. Give *one* reason why the results are likely to be biased.   WM77

7. Give *one* reason why a survey on incomes carried out by phoning telephone subscribers at random is likely to give biased results.   WM76

8. Give *one* reason why a postal questionnaire is likely to give biased results.   WM75

9. You are asked to conduct a survey into the mode of transport used by pupils in travelling to and from your school. Design a questionnaire suitable for this purpose.

This questionnaire is to be given to a random sample of 50 pupils in your school. Describe precisely how you would select this random sample.

WM78

10. You have been asked to conduct an enquiry into the sporting activities of pupils aged 11–18 at a certain mixed school. The enquiry is to include sports in which the pupils take part themselves, and also sports which they watch (including watching television). The enquiry is intended to investigate how the time spent on these activities varies according to age and sex.

Design a questionnaire suitable for obtaining the information which you require.

WM75

11. (a) What are the main principles to be borne in mind when the questions for a questionnaire are being compiled?

(b) Write out a questionnaire that could be used to find out the television viewing habits of the pupils in your school.

(c) Before giving the questionnaire to pupils in your school, what precautions would you take to ensure that the questions that you have just drafted are suitable?

(d) Explain the difference between a 'biased' sample and a 'representative' sample. Describe how you would select a sample of children from your school to answer the questionnaire so that the sample was reasonably representative of all the pupils in the school.

(e) Outline the ways in which the answers to the questionnaire would be recorded, tabulated and analysed.

(f) Suggest an alternative method that could have been used, instead of a questionnaire, to collect data concerning the television viewing habits of the children in your school.

M76

12. (a) Describe in detail how you would collect, record and present information about:

(i) the amount of pocket money received each week by the pupils in your class at school

(ii) the television viewing habits of *all* the pupils in your school.

To what uses might each of these pieces of information be put?

(b) (i) Explain briefly what is meant by 'bias' in sampling.

(ii) How might bias occur in an investigation into the leisure activities of the people of the town of Hillbridge if the information was collected by interviewing people in the main shopping centre? What might be a better method of carrying out such an investigation?

M78

13. State *six* general rules to bear in mind when designing a questionnaire which members of the public will be asked to fill in.

Manufacturers of the Woof-Bang car wanted to know what features of a motor car potential buyers consider important. They decided to ask members of the public to fill in a questionnaire. At a meeting of the committee appointed to run the scheme, the following questions were amongst those suggested for inclusion in the questionnaire.

(*a*) How much money do you earn?

(*b*) What do you think a car should look like?

(*c*) Do you consider child-proof locks are necessary on cars?

State with reasons which (if any) of these questions you think should be included, and which (if any) should not be included.

List *three* further questions which you think could be asked.   EA78

14. A survey of the living accommodation available per family is to be made in a city of 1 million inhabitants. 500 families are to be questioned.

(*a*) State clearly what information should be obtained.

(*b*) List 4 questions which should be asked.

(*c*) Describe precisely how you would select the sample of 500 families.
                                                                    WM76

15. Describe any statistical enquiry in which you have taken part, or in which you would like to take part.

Your answer should include:

(*a*) the object of the enquiry,

(*b*) the methods for getting information,

(*c*) the precautions needed to avoid bias,

(*d*) the method for tabulating the results.   WM77

APPENDIX

Data collection projects which could be undertaken by pupils with a view to the subsequent processing of the collected data.

Personal details of pupils
Ages, weights, heights, sex, hair colouration, shoe size, birthdays, mode of transport to school, size of family, how the day is spent, favourite foods, drinks, amount of pocket money, christian names, bed times, type of holiday preferred, eye colouration.

Pupil interests
Hobbies, pets, intended careers, sports, type of T.V./radio programmes preferred, musical tastes.

School based activities
Time spent on various subjects, both in class and on homework, favourite subjects, analysis of public examination marks and/or grades, views on school dinners, views on school rules/organisation, analysis of school absences, survey of time of arrival of pupils at school.

Sport
Favourite sports, number of goals scored by the local football team each Saturday over a given period of time, number of goals scored by all of the football league clubs each Saturday over a given period of time, times taken by a group of pupils to run 100 yards, batting scores of a chosen group of batsmen during the cricket season.

Book survey
Number of words per sentence in different books by the same author, number of words per sentence in books by different authors, number of letters per word, type of books read by pupils, authors read by pupils.

Traffic survey
Number of each type of vehicle passing, colour of vehicles, passengers per vehicle, make of vehicle, year of manufacture of vehicle (by registration number).

Weather survey
Average daily temperature, daily rainfall, daily barometric pressures, daily relative humidity, daily cloud cover, daily hours of sunshine, daily maximum and minimum temperature.

Biological survey
Number of peas per pod, number of roses per bush etc. Growth of similar plants under varying conditions, e.g. use of different amounts of water, light, fertiliser.

Newspaper survey
Type of newspaper read by parents.
Analysis of use of space by different newspapers.
Analysis of diagrams published in newspapers.

Housing survey
Types of housing in the area of the school.
Price of houses in the area (from newspaper advertisements).

Consumer survey
Cost of second hand cars (from newspaper advertisements).
Weights of sugar bags, flour bags, etc. compared with marked weights.
Amount of petrol used each week by parents.
Amount of petrol sold each week by local filling station.

Published Statistics survey
Analysis of statistics published in newspapers, periodicals etc.

Answers

Exercise 2a, page 16

1.

No.	F
0	1
1	2
2	3
3	5
4	0
5	3
6	7
7	4
8	3
9	2

2.

No.	F
0	1
1	2
2	3
3	5
4	7
5	11
6	7
7	3
8	4
9	2

3.

No.	F
0	3
1	5
2	8
3	11
4	16
5	6
6	4
7	4
8	2
9	1

4.

No.	F
0	2
1	3
2	9
3	14
4	20
5	23
6	15
7	7
8	5
9	2

5.

Letter	F
A	1
B	1
C	3
D	2
E	4
F	2
G	0
H	5
I	0
J	5
K	8
L	11
M	11
N	20
O	0
P	20
Q	0
R	28
S	17
T	12

6. (*a*)

No.	F
1	0
2	9
3	27
4	13
5	14
6	13
7	7
8	1
9	5
10	1

6. (*b*)

No.	F
0	1
1	43
2	32
3	10
4	4
5	0

6. (*c*)

Vowel	F
a	32
e	57
i	22
o	32
u	10

6. (*d*)

Letter	F
A	32
B	7
C	7
D	15
E	57
F	17
G	11
H	26
I	22
J	0
K	3
L	17
M	3
N	28
O	32
P	10
Q	0
R	25
S	31
T	37
U	10
V	4
W	11
X	0
Y	6
Z	0

8.

CF	≤
1	0
3	1
6	2
11	3
11	4
14	5
21	6
25	7
28	8
30	9

9.

CF	≥
2	9
5	8
9	7
16	6
19	5
19	4
24	3
27	2
29	1
30	0

10.

CF	≤
1	0
3	1
6	2
11	3
18	4
29	5
36	6
39	7
43	8
45	9

11.

CF	≥
2	9
6	8
9	7
16	6
27	5
34	4
39	3
42	2
44	1
45	0

12.

CF	≤
3	0
8	1
16	2
27	3
43	4
49	5
53	6
57	7
59	8
60	9

13.

CF	≥
1	9
3	8
7	7
11	6
17	5
33	4
44	3
52	2
57	1
60	0

14.

CF	≤
2	0
5	1
14	2
28	3
48	4
71	5
86	6
93	7
98	8
100	9

15.

CF	≥
2	9
7	8
14	7
29	6
52	5
72	4
86	3
95	2
98	1
100	0

Exercise 2b, page 21

1. (a) 100, (b) 10, (c) 9, (d) 50, (e) 79, (f) 14.5, 44.5, 64.5, (g) 88.   **2.** (a) 120p, (b) 34.5, 49.5, 69.5, 94.5, 119.5, 134.5, 145, (c) 39, 79, 109, 150, (d) 30, 80, 110, 140, (e) 1st and 6th; 2nd, 3rd and 5th, (f) 4th.   **7.** (a) D, (b) C, (c) C, (d) C, (e) D, (f) C, (g) D, (h) C.

Exercise 2c, page 29

1.

Class	F
1 10	1
11–20	2
21–30	5
31–40	7
41–50	10
51–60	6
61–70	3
71–80	3
81–90	2
91–100	1

2.

Class	F
0– 9	1
10–19	1
20–29	4
30–39	8
40–49	10
50–59	15
60–69	12
70–79	5
80–89	3
90–99	1

3. (a)

CF	≤
1	10.5
3	20.5
8	30.5
15	40.5
25	50.5
31	60.5
34	70.5
37	80.5
39	90.5
40	100.5

3. (b)

CF	≥
1	90.5
3	80.5
6	70.5
9	60.5
15	50.5
25	40.5
32	30.5
37	20.5
39	10.5
40	0.5

4. (*a*)

CF	≤
1	9.5
2	19.5
6	29.5
14	39.5
24	49.5
39	59.5
51	69.5
56	79.5
59	89.5
60	99.5

4. (*b*)

CF	≥
1	89.5
4	79.5
9	69.5
21	59.5
36	49.5
46	39.5
54	29.5
58	19.5
59	9.5
60	−0.5

5. 20.5, 40.5, 10.   **6.** 39.5, 69.5, 10.

Miscellaneous Exercises 2, page 30

1.

No.	F
0	1
1	1
2	1
3	2
4	3
5	5
6	4
7	3
8	0
9	0

2.

No.	F.
0	0
1	2
2	3
3	6
4	10
5	14
6	9
7	8
8	4
9	3
10	1

3.

No.	F
0	1
1	5
2	21
3	11
4	11
5	17
6	6
7	13
8	5
9	2
10	4
11	2
12	2

4.

No.	F
0	0
1	1
2	2
3	4
4	8
5	9
6	10
7	9
8	4
9	2
10	1

5.

CF	≤
1	0.5
2	1.5
3	2.5
5	3.5
8	4.5
13	5.5
17	6.5
20	7.5
20	8.5
20	9.5

6.

CF	≤
0	0.5
2	1.5
5	2.5
11	3.5
21	4.5
35	5.5
44	6.5
52	7.5
56	8.5
59	9.5
60	10.5

7.

CF	≤
1	0.5
6	1.5
27	2.5
38	3.5
49	4.5
66	5.5
72	6.5
85	7.5
90	8.5
92	9.5
96	10.5
98	11.5
100	12.5

8.

CF	≤
0	0.5
1	1.5
3	2.5
7	3.5
15	4.5
24	5.5
34	6.5
43	7.5
47	8.5
49	9.5
50	10.5

9.

Class	F
1–3	12
4–6	12
7–9	16

12.

Class	F
0–9	1
10–19	0
20–29	1
30–39	3
40–49	2
50–59	5
60–69	4
70–79	0
80–89	3
90–99	1

13. (a) 29, (b) 10, (c) 49.5, (d) 59.5, (e) 10.

14.

Class	F
0–20	3
21–40	10
41–50	28
51–60	13
61–70	4
71–100	2

10.

Class	F
0–1	10
2–3	7
4–5	14
6–7	9
8–9	10

15. (a) 40.5, (b) 51, (c) 20, (d) 10, (e) 6th, i.e. 71–100.   **16.** 999.5, 1199.5.

17.

CF	≤
12	3.5
24	6.5
40	9.5

19.

CF	≤
1	9.5
1	19.5
2	29.5
5	39.5
7	49.5
12	59.5
16	69.5
16	79.5
19	89.5
20	99.5

20.

CF	≤
3	20.5
13	40.5
41	50.5
54	60.5
58	70.5
60	100.5

18.

CF	≤
10	1.5
17	3.5
31	5.5
40	7.5
50	9.5

21.

CF	≤
1	1.5
5	2.5
10	3.5
25	4.5
50	5.5
82	6.5
104	7.5
112	8.5
115	9.5
117	10.5

22.

CF	≤
5	55.5
13	60.5
28	65.5
45	70.5
55	75.5
62	80.5
65	85.5

Exercise 3a, page 37

1. 1 page ≡ 10°, News 110°, Sport 50°, Adverts 80°, T.V./Radio 20°, Fashions 30°, Ed./Letters 20°, Finance 10°, Features 40°.   **2.** 1 rosebush ≡ 15°, Peace 105°, Queen Elizabeth 75°, Iceberg 30°, Masquerade 60°, Super Star 90°.   **3.** 1 hour ≡ 24°, Films 96°, Doc./Curr. Aff. 48°, Children 60°, Comedy 36°, News 24°, Serials/Plays 72°, Sport 24°.
4. 1 pupil ≡ 3°, Thrillers 60°, Rom. Novels 120°, Sci.Fi. 45°, Technical 30°, Det.Stories 75°, Classics 30°.   **5.** 1 girl ≡ 1.5°, Earrings 45°, Pendants 60°, Rings 75°, Bracelets 90°, Chokers 30°, Necklaces 45°, Brooches 15°.

Miscellaneous Exercises 3, page 60

1. 1 pupil ≡ 12°, Abroad 168°, Seaside 72°, Other 96°, None 24°.   **2.** £1,000 ≡ 3°, Food 180°, Housing 60°, Clothing 45°, Tob./Drink 51°, Entertainment 24°.   **3.** 1 hour ≡ 15°, English 45°, French 45°, Maths 60°, Physics 37.5°, Chemistry 37.5°, Others 135°.   **4.** 1% ≡ 3.6°, Work 198°, Elsewhere 162°.   **5.** (a) 1 hour ≡ 15°, Sleep 135°, Work 105°, Play 45°, Meals 37.5°, Other 37.5°.   **6.** (a) £1,000,000.   **7.** (a) 1 car ≡ 3°, France 135°, Germany 54°, Holland 30°, Spain 81°, Italy 60°.   **8.** (b) 1 girl ≡ 15°, Sport 90°, 6 girls named Sport.   **9.** (a) 1 child ≡ 1.5°, Bus 132°, Cycle 96°, Walk 60°, Car 48°, Train 24°.   **10.** (b) 1% ≡ 3.6°, Sickness 136.8°, (c) 60 pupils.   **14.** 600 nails.   **15.** 18 cm.   **16.** 920 nails.   **17.** (a) 50°, (b) 1° ≡ £0.2625 million. Social services £7.35 million. (c) 60%.   **18.** (a) 1 hat ≡ 10,000 civil servants, (b) 35,000, (c) 2.8 hats.   **19.** (a) 4 mins, (b) 6 hours, (c) 8 hours.   **20.** (a) 1° ≡ 0.88p, Writer's fee 36p, (b) Publishing 170°, Cost of publishing 151p.   **21.** (a) x = 100°, (b) 63 people, (c) 28° and 42°.   **22.** (a) 2 in group D, (b) 25% in group A.   **23.** 40.   **24.** (a) Sun 400. Mon 50. Tue 75, Wed 100. Thur 150. Fri 250. Sat 375, (b) 1 vote ≡ 0.1°, Gooseberry 70°, Blackcurrant 20°, Apple 50°, Blackberry 100°, Cherry 120°.   **28.** 1 minute ≡ 8°, Fill 64°, Wash 120°, Empty 64°, Rinse 80°, Spin 32°.   **29.** (a) (ii) 4½, (b) 1 boy ≡ 3°, Soccer 138°, Rugby 99°, Cricket 75°, Hockey 48°.   **30.** (b) (iii) 1 million ≡ 30°, Home 240°, Africa 90°, Australia 30°.   **31.** (a) (i) 1 km ≡ 2°, Motorway 150°, Dual 100°, 3 lane A 40°, 2 lane A 50°, B 20°.

Exercise 4a, page 77

1. (a)

No.	F
1	1
2	1
3	3
4	1
5	1
6	2
7	1

mode = 3

1. (b)

No.	F
1	1
2	1
3	1
4	0
5	4
6	0
7	1
8	1
9	1

mode = 5

1. (c)

No.	F
1	1
2	1
3	3
4	0
5	4
6	0
7	0
8	1

mode = 5

1. (d)

No.	F
1	1
2	0
3	0
4	4
5	1
6	3
7	1
8	1

mode = 4

1. (e)

No.	F
0	1
1	0
2	3
3	2
4	0
5	1
6	0
7	1
8	0
9	1

mode = 2

1(f)

No.	F
1	0
2	1
3	1
4	3
5	5
6	1
7	0
8	1
9	1

mode = 5

1. (g)

No.	F
1	1
2	2
3	6
4	1
5	0
6	1
7	2
8	0
9	1

mode = 3

1. (h)

No.	F
0	2
1	1
2	1
3	1
4	5
5	2

mode = 4

2.

No.	F
1	2
2	1
3	4
4	7
5	4
6	4
7	4
8	2
9	2

mode = 4

3.

No.	F
1	4
2	0
3	0
4	4
5	12
6	9
7	5
8	3
9	3

mode = 5

4.

No.	F
1	2
2	5
3	10
4	6
5	4
6	2
7	0
8	2
9	2

mode = 3

5.

No.	F
1	2
2	4
3	8
4	16
5	7
6	4
7	3
8	1

mode = 4 fillings

6.

No.	F
0	5
1	3
2	8
3	17
4	26
5	10
6	5
7	1

mode = 4 eggs

7.

No.	F
1	1
2	1
3	3
4	5
5	7
6	8
7	12
8	6
9	3
10	3
11	1
12	0
13	1
14	0
15	1

mode = 7 gallons

8.

No.	F
20	2
21	1
22	2
23	2
24	1
25	3
26	4
27	7
28	3
29	1
30	1
31	3

mode = 27°C

9. 251 pages. **10.** Bimodal, 60 and 47 stitches. **11.** 6 metres. **12.** 70 strokes.
13. 4.0 metres.

Exercise 4b, page 82

1.

Class	F
1–10	1
11–20	7
21–30	5
31–40	6
41–50	10
51–60	6
61–70	8
71–80	4
81–90	2
91–100	1

modal interval
41–50 mins.

2.

Class	F
45–59	1
60–74	8
75–89	10
90–104	25
105–119	12
120–134	3
135–149	1
150–164	0

modal interval
90–104 mins.

3. 41–50 runs.
4. 1,000–1,099 bricks.
5. 4–5 s.
6. 18,000–19,000 pounds.
7. 400–599 g.
8. 3–5 mins.
10. 45.5 mins, 97.75 mins, 44.75 runs, 1049.5 bricks, 4.39 s, £18,550, 511.5 g, 4.3 mins.

Exercise 4c, page 91

1. (*a*) 3.5. (*b*) 5. (*c*) 4. (*d*) 5. (*e*) 3. (*f*) 5. (*g*) 3. (*h*) 4. **2.** 5. **3.** 5½. **4.** 3. **5.** 4 fillings. **6.** 4 eggs. **7.** 7 gallons. **8.** 26.5°C. **9.** 252 pages. **10.** 50.5 stitches. **11.** 8 metres. **12.** 71.5 strokes. **13.** 4.0 metres.

Exercise 4d, page 94

1. 46.5 mins. **2.** 97 mins. **3.** 48 runs. **4.** 1,050 bricks. **5.** 4.3 s. **6.** £18,800. **7.** 590 g. **8.** 6.6 mins.

Exercise 4e, page 98

1. (*a*) 4. (*b*) 5. (*c*) 4. (*d*) 5. (*e*) 3.6. (*f*) 5. (*g*) 4. (*h*) 3. **2.** 5. **3.** 5.5. **4.** 4. **5.** 4.13 fillings. **6.** 3.48 eggs. **7.** 6.60 gallons. **8.** 25.97°C. **9.** 252.65 pages. **10.** 51.475 stitches. **11.** 9.28 metres. **12.** 71.71 strokes. **13.** 4.205 metres.

Exercise 4f, page 100

1. 46.7 mins. **2.** 95 mins. **3.** 51.99 runs. **4.** 1,052.36 bricks. **5.** 4.42 s. **6.** £19,100. **7.** 645.5 g. **8.** 7.9 mins. *Note.* In question 3 the mid-value of the class 0–10 is 5.

Miscellaneous Exercises 4, page 102

1. 45. **2.** 5. **3.** 4. **4.** 3. **5.** 9. **6.** (*a*) 41 points, (*b*) 43 points. **7.** 42.5. **8.** mean = 22, median = 21, mode = 19. **9.** (i) 4, (ii) 3, (iii) 5. **10.** (*a*) mode = 6, median = 5. (*b*) mean = 5.2. (*c*) 44%. **11.** 123. **12.** (*a*) (i) 23.6 hours, (ii) 25 hours. (*b*) 28 hours. **13.** mean = 9.4, median = 9.3, mode = 9.1. **14.** (*a*) 21.25, 26.65, 28.50, 33.40, 43.75. (*b*) (i) 28.50 (ii) 30.71. **15.** 52. **16.** 9. **17.** 9.904. **18.** 179.7. **19.** 250.1. **20.** (*a*) 178 cm, (*b*) 177 cm, (*c*) 176 cm, (*d*) 35%. **21.** 5.74 cars. **22.** (*a*) 500, (*b*) (i) 82, (ii) 322, (*c*) (i) 50–59, (ii) 40–49. **22.** (*d*) (i) 30, 231, 157, 82, (ii) 49, 28.

23.

Class	F
35–39	3
40–44	2
45–49	5
50–54	6
55–59	4
60–64	0
65–69	2

modal class 50–54

24. (*a*)

Class	F
135–139	2
140–144	4
145–149	11
150–154	9
155–159	5
160–164	1

modal class 145–149

24. (*b*) (i) 12 pupils ⩾ 75p, 9 pupils ⩽ 50p, (ii) median = 70p, range = 120p.
25. (*a*) (i) 14, (ii) 13, (iii) 12,

(*b*)

Wages	F
40–49	1
50–59	3
60–69	11
70–79	8
80–89	7

modal class 60–69.

26.

Mark	B	G	All
10	1	0	1
9	0	1	1
8	1	1	2
7	1	2	3
6	3	4	7
5	4	4	8
4	3	0	3
3	1	1	2
2	1	0	1
1	0	1	1
0	1	0	1

26. (*a*) 10, (*b*) 8, (*c*) 5, (*d*) (i) 5.3, (ii) 5, (iii) 5.

27. (*a*)

Time	F	T × F
24	1	24
25	2	50
26	4	104
27	3	81
28	0	0
29	2	58
30	1	30
31	3	93
32	4	128
Totals	20	568

27. (*b*) 28.4 s. (*c*) 28 s. **28.** (*a*) 32. (*b*) 5. (*c*) 5. **29.** 0.662 passengers. **30.** £4.15.
31. 4.82.

32.

Class	F
1–3	18
4–6	18
7–9	14

33. 4.76. **34.** Yes since Mid-values are used in 33. **35.** (*a*) sum = 102, mean = 17, (*b*)
117. **36.** 11. **37.** 84p. **38.** 9.2 hours. **39.** (*a*) 54, (*b*) 40.6 kg. **40.** (*a*) 5.6 cm.
41. 174 cm. **42.** 4.8. **43.** (*a*) 18.25, (*b*) 17.5. **44.** (*a*) 13.44 blooms, (*b*) 55, 80, 90,
97, (*c*) 12.25 blooms. **45.** (*a*) cum. freqs. 3, 12, 24, 42, 62, 92, 136, 174, 201, 221, 234,
243, 248, 250, (*b*) (i) 34, (ii) 62%. **46.** mode.

49. (*a*)

Class	F	CF	≤
0–10	1	1	10.5
11–20	2	3	20.5
21–30	4	7	30.5
31–40	9	16	40.5
41–50	14	30	50.5
51–60	15	45	60.5
61–70	8	53	70.5
71–80	1	54	80.5
81–90	1	55	90.5
91–100	0	55	100.5

50. (*a*)

Class	F
0–9	10
10–19	21
20–29	18
30–39	16
40–49	8
50–59	4
60–69	1
70–79	1
80–89	1

50. (*c*) median is in 20–29.

49. (*b*) median = 49.

51. (*a*) 828 kg, (*b*) 456 kg, (*c*) 64.2 kg.

52.

Class	F	CF	≤
0–10	3	3	10.5
11–20	5	8	20.5
21–30	10	18	30.5
31–40	15	33	40.5
41–50	15	48	50.5
51–60	13	61	60.5
61–70	9	70	70.5
71–80	6	76	80.5
81–90	3	79	90.5
91–100	1	80	100.5

52. (*b*) median = 45.
53. £46.87.
54. 37.9 points.

Exercise 5a, page 124

Mean denoted by M, mean deviation denoted by MD
1. (*a*) M = 7, MD = 2.86, (*b*) M = 7, MD = 4, (*c*) M = 17, MD = 4, (*d*) M = 14, MD
= 5.5, (*e*) M = 34, MD = 5.5. **2.** M = 4, MD = 1.67. **3.** M = 11, MD = 3.28.
4. M = 6, MD = 1.4. **5.** M. = 15.5, MD = 1.8. **6.** M = 26, MD = 11.45. **7.** M =
49, MD = 13.32. **8.** M = 31, MD = 3.33. **9.** M. = 47.5, MD = 14.32. **10.** M =
49.5, MD = 17.75.

Exercise 5b, page 131

Means as given in answers to Ex. 5a
1. (*a*) 3.46, (*b*) 4.84, (*c*) 4.84, (*d*) 6.73, (*e*) 6.73. **2.** 2.18. **3.** 4.2. **4.** 1.91.
5. 2.25. **6.** 13.52. **7.** 17.26. **8.** 4.09. **9.** 18.11. **10.** 22.14.

Miscellaneous Exercises 5, page 138

1. (i) £25, (ii) 30 people, (iii) £15.50,

(iv)

CF	F
10	10
30	20
66	36
89	23
100	11

3. (*a*) 148, (*b*) 39, (*c*) (i) 50.5, (ii) 37.5. **4.** (i) Range A = 0.6 kg, Range B = 0.6 kg,
Mean A = 15.007 kg, Mean B = 14.998 kg, (iii) Mode A = 15.0 kg, Mode B = 15.0 kg,
(iv) Inter A = 0.2 kg, Inter B = 0.2 kg.

5.

CF	$\leqslant$
18	10.5
57	20.5
150	30.5
277	40.5
480	50.5
690	60.5
861	70.5
949	80.5
980	90.5
1000	100.5

(*b*) (i) 51, (ii) 61,
(iii) 200,
(iv) 37th.

6.

CF	$\leqslant$
5	9.5
15	19.5
60	29.5
125	39.5
230	49.5
350	59.5
425	69.5
465	79.5
490	89.5
500	99.5

(*a*) 51, (*b*) 75,
(*c*) 35, (*d*) 26th.

7.

CF	$\leqslant$
22	14.5
68	29.5
154	44.5
262	59.5
409	74.5
508	89.5
568	104.5
600	119.5

(i) 64 apprentices,
(ii) 49 apprentices,
(iii) 190 factories.

8.

CF	$\leqslant$
2	30.5
7	31.5
25	32.5
69	33.5
130	34.5
175	35.5
195	36.5
199	37.5
200	38.5

(iii) 34 gm,
(iv) 1.7 gm,
(v) 34.35 gm,
(vi) 7%.

9.

CF	≤
0	20
15	30
57	40
122	50
214	60
289	70
356	80
393	90
400	100

(c) (i) 244 men, (ii) £74, (iii) £37. **11.** (b) median 138 cm, LQ = 134 cm, UQ = 142 cm, SIQR = 4 cm. **12.** (b) (i) UQ = 81 kg, (ii) LQ = 75 kg, (iii) SIQR = 3 kg. **13.** (b) (i) UQ = 61 kg, (ii) LQ = 55 kg, (iii) SIQR = 3 kg.

14. (i)

≤	CF
60	1
70	4
80	16
90	39
100	64
110	86
120	96
130	99
140	100

(iii) 94.5, (iv) UQ = 105, LQ = 84, (v) IQR = 21, (vi) Normal. **15.** 2. **16.** 2. **17.** 20. **18.** 20. **19.** 4. **20.** 20. **21.** 20. **22.** 4. **23.** 4. **24.** 20. **25.** (a) 47, (b) 15.91. **26.** a) 0.4, yes, (b) mean 16–1, median 16–1, mode 15–11. **27.** mean = 51, variance = 286.5, standard deviation = 16.92. **28.** (i) 12 hours, (ii) 12.25 hours, (iii) standard deviation = 5.55 hours. **29** (a) 19°C, (b) 1.77°C. **30.** mean = 32, variance = 14.4, standard deviation = 3.80. **31.** mean = 6 half days, standard deviation = 1.41 half days.

32. (a)

$$
\begin{array}{ccccccc}
 & 1 & 4 & 6 & 4 & 1 & \\
1 & 5 & 10 & 10 & 5 & 1 & \\
1 & 6 & 15 & 20 & 15 & 6 & 1 \\
\end{array}
$$

32. (b) mean = 18, standard deviation = 4.69. **33.** (a) (i) 45 kg, (ii) 50 kg, (iii) 46 kg. (b) mean = 7, standard deviation = 4.24. **34.** 7.2 m. **35.** 2.87. **36.** 5.74.

Miscellaneous Exercises 6, page 164

1. (a) (i) $\frac{3}{100}$. **2.** (a) (i) $\frac{3}{5}$, (ii) $\frac{2}{25}$. **3.** (a) $\frac{3}{4}$, (b) $\frac{1}{16}$. **4.** (a) (ii) 75, (b) (ii) 25.125. **5.** $\frac{1}{10}$. **6.** $\frac{9}{10}$. **7.** $\frac{81}{100}$. **8.** Nothing. **9.** $\frac{1}{13}$, **10.** (a) $\frac{4}{11}$, (b) $\frac{10}{11}$. **11.** $\frac{1}{3}$. **12.** (a) $\frac{5}{13}$, (b) $\frac{1}{4}$. **13.** (a) $\frac{2}{3}$, (b) $\frac{4}{9}$. **14.** (a) $\frac{1}{6}$, (b) (i) $\frac{1}{36}$, (ii) $\frac{1}{12}$, (iii) $\frac{11}{12}$. **15.** (a) 36, (b) (i) $\frac{1}{9}$, (ii) $\frac{7}{12}$, (iii) $\frac{5}{12}$. **16.** $\frac{1}{18}$. **17.** (b) (i) $\frac{1}{5}$, (ii) $\frac{1}{10}$, (c) $\frac{3}{5}$. **18.** (a) 24, (b) $\frac{3}{10}$. **19.** (a) (i) H1, H2, H3, H4, H5. H6. T1. T2. T3. T4. T5. T6. (ii) (A) $\frac{1}{4}$, (B) $\frac{1}{3}$. (b) (i) $\frac{1}{2}$, (ii) $\frac{2}{3}$. **20.** (b) (i) 36, (ii)

$\frac{1}{9}$, $\frac{1}{12}$, $\frac{1}{6}$, $\frac{5}{12}$, (iii) $\frac{11}{36}$, (c) $\frac{1}{144}$. **21.** (b) $\frac{1}{9}$, (c) $\frac{5}{9}$, (d) $\frac{2}{27}$. **22.** (a) SCT, SCI, SSaT, SSaI, FCT, FCI, FSaT, FSaI, (b) $\frac{1}{4}$. **23.** $\frac{2}{5}$. **24.** $\frac{8}{15}$. **25.** (a) (i) BBB, BBG, BGB, BGG, GBB, GBG, GGB, GGG, (ii) 16, (b) (i) $\frac{3}{8}$, (ii) $\frac{7}{8}$. **26.** (a) (i) $\frac{1}{8}$, (ii) $\frac{2}{3}$, (iii) $\frac{1}{8}$, (b) (i) $\frac{1}{36}$, (ii) $\frac{1}{9}$, (iii) $\frac{1}{9}$, (c) $\frac{1}{54}$, (d) 10p. **27.** (a) 408, (b) 1 pupil $\equiv$ 0.8824°, Maths 105°, (c) $\frac{17}{24}$. **28.** (a) (i) $\frac{1}{40}$, (ii) $\frac{1}{4}$, (iii) 7, (iv) 22, (b) Green 1, Red 1, Blue 3, Yellow 5. **29.** (a) (i) $\frac{1}{2}$, (ii) $\frac{1}{10}$, (iii) $\frac{3}{10}$, (iv) $\frac{3}{10}$, (v) $\frac{1}{2}$, (c) 20, 80, 120, 80, 20.

Miscellaneous Exercises 7, page 182

2. (6, 5). **3.** (66.7, 59.2). **4.** (a) $\frac{3}{10}$, (b) (C, 49.3), (ii) Fairly good positive. **5.** (a) (i) 1.41, (ii) Dispersion, (b) (67, 44.4), (ii) High positive. **6.** (175, 155), F tallest, J shortest. **7.** (i) (30.3, 26.8), (ii) High positive, (iii) 25. **8.** (a) (43.7, 53.6), (c), (i) 70, (ii) 22, (d) High positive. **9.** (a) (10.07, 306.7), (c) High positive, (d) (i) 375 litres, (ii) 12°C. **10.** (10.25, 4,374), (a) 32.5 words, (b) 46.2 words. **11.** (4.5, 71.8), (b) (i) 56 cm, (ii) 3.7 kg, (iii) 82.2 cm. **12.** (a) (28.6, 23.6), (b) (i) 16, (ii) 24, (iii) 15.5 and 14.5. **13.** (a) (43.7, 36.8), (iii) 43. **14.** (a) Exports £1,337M, Imports £1,768M, (137,568), (c) fairly low negative.

Exercise 8a, page 192

1. 6, 10, 10, 12, 16, 14, 10, 12, 18, 16, 10. **2.** 9, 8.7, 8.3, 8, 7.7, 7.3, 7. **3.** 10.75, 10, 9, 8, 7.25, 6.75, 6.5, 6, 5.5. **4.** 38.2, 37, 35.8, 31.2, 30.4, 26.8, 25.6, 22.2. **5.** 25.17, 26.17, 27.17, 28.17, 27.7, 27, 26.17, 26.3, 27, 28.

Miscellaneous Exercise 8, page 197

2. 10.2, 9.6, 8.2. **3.** 13, 13.2, 13, **4.** 13. **5.** 19. **6.** 7. **7.** (ii) 396, 426, 452, 480, 500, 514, 524, 534, 548, 556, 564. **8.** (ii) 6.1, 6.47, 6.3, 6.03, 7.07, 7.37, 7.03, 6.9, 7.73, 8.23, (iv) 7.6. **9.** (b) 13.275, 13.05, 12.875, 12.625, 12.275, 12.225, 11.85, 11.5, 11.3, (d) 14.0%. **10.** (b) 813.25, 832.25, 850.5, 866, 885, 907.5, 932.75, 953.25, 968. **11.** (b) 244, 258, 265, 267, 271, 281, 289, 299, 309, 320. **12.** (b) 30, 32, 34, 36, 38, 40, 42, 42, 42, 44, 44, 44.3, 48.3. **13.** (ii) 4, 4.25, 4.5, 4.75, 4.5, 4.5, 4.75, 5.25, 5.75. **14.** (b) 31.5, 32.775, 33.925, 35.075, 36.425, 38, 39.375, 40.8, 42.425, (d) 49.9. **15.** 80.5, 81.25, 81.25, 81, 81, 81.25, 82.25, 80.75, 83, 84, estimate 84.

Miscellaneous Exercises 9, page 207

1. 4, 7, 1, 10, 7, 5, 1, 9, 3, 7. **2.** R must lie between −1 and +1. **3.** −1, **4.** (i) 0.491, (ii) (44.4, 42.2), Maths 46. **5.** (a) 0.682, (c) (48.6, 56.6), (e) 68. **6.** (b) 0.8007. **7.** (b) 0.6875, **8.** (b) 0.9598. **9.** −0.9515.

Exercise 10a, page 217

1. 34. **2.** 25.75. **3.** 529. **4.** 24.8. **5.** 101.

Exercise 10b, page 223

1. 46.5.   **2.** 45.   **3.** 50.5.   **4.** 360.5.   **5.** 29.25.

Exercise 10c, page 232

1. (*a*) 10.83, (*b*) 7.84, (*c*) 1.69, (*d*) 1.38, (*e*) 4.55.   **2.** (*a*) 21.22, (*b*) 5.4, (*c*) 15.13, (*d*) 140.2, (*e*) 9.82.

Miscellaneous Exercises 10, page 232

1. (*a*) 11.93 s, (*b*) 40.55, **2.** (*a*) 40.5, (*b*) (i) range = 12, median = 11, mean = 10.5, (ii) range = 15, median = 22, mean = 22.45.   **3.** (*a*) freqs. 1, 1, 3, 4, 6, 6, 5, 2, 7, 6, 6, 3, 2, 5, 2, 1, (*b*) 40.67.   **4.** (*a*) 0.5, 1.5, 2.5, 4, 6.5, 11.5, (*b*) Mean = 2.4 tons, standard deviation = 2.8 tons.   **5.** mean = 47, standard deviation = 15.33.   **6.** mean = 54.15, standard deviation = 6.48.   **7.** mean = 28.42, standard deviation = 5.85.   **8.** (*a*) 1017.5 hours, (*b*) 191 hours.   **9.** (i) 67.82 in, (ii) 3.18 in.   **10.** (i) 34.28 years, (ii) 21.89 years. **11.** (i) 169 lb, (ii) 21.9 lb.

Miscellaneous Exercises 11, page 244

1. 64,   **2.** 51.   **3.** (*a*) (i) Mean A = 62, Mean B = 62, Mean C = 62, (ii) W Mean A = 63.5, W Mean B = 62.5, W Mean C = 65, (*c*) (i) C.   **4.** (*a*) Mean P = 70, Mean Q = 70, (*b*) W Mean P = 66, W Mean Q = 72.8, Diff. P = −4, Diff. Q = +2.8.   **5.** 40. **6.** 80.   **8.** 200.   **9.** £48.   **10.** (*a*) For 1935, 66, 4.5, 91, For 1975, 605, 57, 833,(*b*) £7.42, (*c*) 815.38%, Price relative 915.38.   **11.** (*a*) 121.92, (*b*) 3.32.   **12.** 131. **13** (*a*) 129.4, (*b*) 6.

Miscellaneous Exercises 12, page 255

1. $\frac{15}{28}$, i.e. C.   **2.** (*b*) $\frac{17}{25}$, (*c*) $\frac{3}{5}$ **3.** (*b*) $\frac{7}{12}$, (*c*) $\frac{1}{14}$.   **4.** (*a*) (i) $\frac{1}{20}$, (ii) $\frac{1}{25}$, (*b*) (i) $\frac{2}{5}$, (iii) 360, 480, 160.   **5.** HTH, HTT, HHH, HHT, TTH, TTT, THH, THT, (*a*) (i) $\frac{1}{8}$, (ii) $\frac{3}{8}$, (*b*) (i) $\frac{1}{24}$, (ii) $\frac{1}{8}$.
6. HHH, HHA, HHD, HAH, HAA, HAD, HDH, HDA, HDD, ⎤
   AHH, AHA, AHD, AAH, AAA, AAD, ADH, ADA, ADD, ⎬ Total 27.
   DHH, DHA, DHD, DAH, DAA, DAD, DDH, DDA, DDD. ⎦
(*b*) (i) $\frac{2}{3}$, $\frac{5}{9}$, $\frac{4}{9}$, (ii) $\frac{1}{27}$, $\frac{1}{9}$, $\frac{4}{9}$.
7. 3 coins, HHH, HHT, HTH, HTT, THH, THT, TTH, TTT, Total 8.
4 coins, HHHH, HHHT, HHTH, HHTT, HTHH, HTHT, HTTH, HTTT, ⎤
         THHH, THHT, THTH, THTT, TTHH, TTHT, TTTH, TTTT. ⎦ Total 16.
(*a*) (i) $\frac{1}{4}$, (ii) $\frac{3}{4}$, (*b*) (i) $\frac{3}{8}$, (ii) $\frac{7}{8}$, (*c*) (i) $\frac{1}{4}$, (ii) $\frac{3}{8}$.
8. (*a*)   1   4   6   4   1
          1   5   10   10   5   1
(*b*) (i) $\frac{1}{8}$, (ii) $\frac{1}{4}$, (iii) $\frac{3}{8}$, (*c*) (i) $\frac{1}{3}$, (ii) $\frac{5}{8}$, (*d*) $\frac{15}{28}$.